AI for Designers

Md Haseen Akhtar · Janakarajan Ramkumar

AI for Designers

 Springer

Md Haseen Akhtar
Indian Institute of Technology Kanpur
Kanpur, Uttar Pradesh, India

Janakarajan Ramkumar
Indian Institute of Technology Kanpur
Kanpur, Uttar Pradesh, India

ISBN 978-981-99-6896-1   ISBN 978-981-99-6897-8   (eBook)
https://doi.org/10.1007/978-981-99-6897-8

This Springer imprint is published by the registered company Springer Nature Singapore Pte Ltd.
The registered company address is: 152 Beach Road, #21-01/04 Gateway East, Singapore 189721,
Singapore

Paper in this product is recyclable.

Future is already here; it is just not deeply imagined of

Md Haseen Akhtar

Dedicated to my mother…and my late father…
May Almighty have mercy on them as they raised me when I was a child…

Md Haseen Akhtar

Preface

Artificial intelligence (AI) has become one of the most transformative technologies of our time, with the potential to reshape industries and revolutionize the way we live and work. As designers, we are uniquely positioned to leverage the power of AI to create new and innovative experiences for users. In recent years, AI has begun to permeate the world of design, with applications ranging from generative art and visual design to predictive analytics and user experience design. AI enables designers to automate repetitive tasks, analyze vast amounts of data, and generate new and creative ideas. But as with any emerging technology, AI can also be intimidating and overwhelming, particularly for designers who may not have a technical background. This book aims to demystify AI and provide designers with the knowledge and skills they need to incorporate AI into their work. Through a combination of theory and practical examples, this book will introduce designers to the fundamentals of AI, explain how it can be applied to design, and offer guidance on how to work effectively with AI tools and technologies. Whether you are a graphic designer, a user experience designer, or a product designer, this book will help you unlock the potential of AI and take your design skills to the next level.

Kanpur, India

Md Haseen Akhtar
Janakarajan Ramkumar

Acknowledgments We would like to acknowledge the support by Priya Vyas in making this idea come to reality and giving us an opportunity to share our thoughts on AI for designers. We would like to thank the authors whose works are credited and referenced in this book.

Welcome from Authors

As a passionate individual in the field of design, I am thrilled to introduce this book on artificial intelligence for designers, written by us. In today's fast-paced world, technology has become an integral part of the design process, and AI is one of the most exciting and transformative technologies in this space. As designers, we have the unique opportunity to leverage the power of AI to create more engaging and compelling experiences for our users. But to do so, we need to have a solid understanding of the fundamentals of AI and how it can be applied in design. This is where this book comes in.

The content of this book has demystified AI and making it accessible to designers of all levels. He covers the key concepts of AI in a clear and concise manner and provides practical examples and case studies to help readers understand how AI can be applied in real-world design scenarios. What sets this book apart, however, is its focus on the creative potential of AI. The book emphasizes the importance of embracing AI as a tool for generating new and innovative ideas, rather than simply automating repetitive tasks. It shows how AI can be used to create generative art, design unique interfaces, and even cocreate with users. As someone who has worked in the field of design for many years, we can attest to the importance of staying up to date with the latest trends and technologies. AI is no exception. We believe that this book will be a valuable resource for designers who are looking to expand their knowledge and skills in the rapidly evolving world of AI. We highly recommend this book to anyone who is interested in learning about the creative potential of AI, and how it can be applied in the field of design. This book has done its job of breaking down complex concepts into easy-to-understand terms, and we are confident that readers will find this book to be an engaging and insightful resource.

Md Haseen Akhtar
Janakarajan Ramkumar

Introduction

As AI continues to evolve, designers must stay up to date with the latest developments and techniques to stay competitive in the field. This book is designed to provide designers with a comprehensive overview of AI and its applications in design. Whether you are a seasoned professional or just starting out in the field, this book will give you the knowledge and tools you need to incorporate AI into your work and take your designs to the next level. The book begins with an introduction to AI for all, to AI for designers and then to more specific fields of design like product design, architecture, visual communication, and interaction design. From there, we delve into the specific applications of AI in design, including generative art, data visualization, and user experience design. Each chapter has been divided into several sections starting with AI for specific topic followed by asking some fundamental questions and answering them through an application of tool, excerpt from a blog, podcast, or a seminar in each addressed field of design. We also explore the ethical implications of AI in design, including issues around bias, transparency, and privacy. Throughout the book, we provide practical examples and case studies to help you understand how AI can be applied in real-world design scenarios. We also provide guidance on how to work effectively with AI tools and technologies, and how to collaborate with AI systems to create more innovative and effective designs. It is our hope that this book will serve as a valuable resource for designers who are looking to expand their knowledge and skills in the rapidly evolving world of AI. Whether you are a graphic designer, a user experience designer, or a product designer, this book will give you the tools you need to stay ahead of the curve and create truly exceptional designs.

Contents

About the Authors

Md Haseen Akhtar is currently Prime Minister Research Fellow at the Department of Design, Indian Institute of Technology Kanpur, Uttar Pradesh. He obtained his Bachelor of Architecture (B.Arch.) from National Institute of Technology, Trichy; Master of Design in Industrial Design from Indian Institute of Technology, Kanpur, and currently pursuing Ph.D. from the Indian Institute of Technology, Kanpur. He was the Gold Medalist of B.Arch. (2015–2020). He is also a recipient of the prestigious Fulbright Nehru Doctoral Research Fellowship 2023–2024 at UC Berkeley, CA, USA, and BIRAC Biotechnology Ignition Grantee Fellow 2022–2023, India. His major areas of research interests include healthcare infrastructure and services design, information design in health care, and architectural design. His research has been published in peer-reviewed international journals, presented at several national and international conferences and has five patents and design copyrights.

Dr. Janakarajan Ramkumar is currently Professor at the Department of Mechanical Engineering and Department of Design, Indian Institute of Technology Kanpur, Uttar Pradesh. He obtained his B.E. (Production Engineering) from National Institute of Technology, Trichy, and M.Tech. and Ph.D. in Mechanical Engineering from Indian Institute of Technology, Madras, Chennai. His major areas of research interests include micro-electric discharge milling, micro-electro-chemical milling, excimer laser micro-machining, abrasive flow finishing, magnetic abrasive finishing, fabrication of composites, and machining of composites. His expertise is also evident in the field of healthcare ranging from primary healthcare innovation, intensive care unit innovation, and biomedical devices innovation with over 100 patents and design copyrights. He has published more than 250 peer-reviewed articles in high-impact journals.

Chapter 1
AI for All

Who is AI? Our Future Friend?

1.1 Artificial Intelligence is for Humanity

This is the "uncommon" made "common" tech in the entire world that has penetrated in all sectors and all walks of life. You use it too. Isn't it?

1.1.1 First Perspective—AI as Replacement for Human

The term artificial intelligence (AI) may sound frightening, but in practice, it is not nearly as frightening as it sounds. When taken to its most inclusive extreme, the definition of artificial intelligence (AI) is a machine that can act using human-style reasoning or perception. There is an underlying anxiety among designers regarding what this could imply for the future of creative work, specifically that AI will be responsible for ideation, design, and presentation. On the other hand, that is not always the case. For a very long time, people have been working towards the creation of machines and technologies that will transform the way things are done. Since the invention of the wheel, humans have been hard at work developing tools and machines that make life and work a little bit simpler. As a direct result of this, not only have we grown as individuals, but also the occupations that we do have changed. This is the same as the situation with AI and design. Therefore, while AI will eventually replace designers, it will only replace the designers working now; it will not replace designers working in the future. AI will evolve into a design collaborator and tool that designers may utilize to satisfy the constantly shifting requirements of the workplace [1].

If there is a discussion on AI, it is all about endless advantages, but there is scarce regarding its limits. Let us start the discussion by knowing the limitations of AI. Emotions have served as a vital means of human self-preservation throughout the course of human history. We are conditioned to utilize feelings such as fear as a means of protecting ourselves from the influences of the outside world. We can detect and decipher the emotions of other people by analyzing their body language, tone of voice, context, and social cues, all of which are based on cultural norms and acquired behaviors. As a result, one of the most difficult problems for AI is to comprehend the nuances of different emotions. During an experiment, study scientist Janelle Shane [2] fed a neural network thousands of pickup lines to demonstrate how flattery can be learned. Here are some memorable lines:

Are you a candle? Because you're so hot of the looks with you.

I want to get my heart with you.

You look like a thing, and I love you.

Even for the most articulate of individuals, flirting conversation may be somewhat challenging. However, the results of her experiment make it abundantly evident that it is difficult to teach intelligent machines to recognize or communicate nuances of nuanced emotional states. The second limitation is producing work autonomously without large sets of data and defined parameters as found by Mario Klingemann [3], machine learning artist, used facial recognition algorithm to paint faces from its pattern-based model. The other limitation is filter biases, identifying biases based on social or ethical consciousness. When Jacky Alcine [4] noticed that his photos app had tagged him and his friend as "gorillas," he had to learn this lesson the hard way. It is difficult to teach morality to machines because humans are unable to objectively convey morality in measurable metrics that make it easy for a computer to process. Artificial intelligence that lacks its own moral and social consciousness can only take datasets without first screening them for any prejudices [1].

What AI is great at is dynamic personalization. For example, consider the most recent occasion in which you were put in contact with an unknown person for the first time. You have most certainly created assumptions about the individual based on their appearance as well as their mannerisms, even if you are aware that you have done so. AI accomplishes the same goal, but it also considers the person's subconscious inclinations. For instance, my Instagram feed is dynamically customized based on the time of day, the posts that I've liked, the amount of time I spend on the app, what my friends are looking at, what events are trending, where I am, and the sort of device that I'm using. The next AI advantage is to handle multiple variables as according to the Huffington Post [5], for human doctors to maintain their level of medical expertise, they are required to spend approximately 160 h per week performing the tedious task of reading research papers. On the other hand, computers are quite good at sifting through dozens or even hundreds of data points all at once. For example, the artificial intelligence powering IBM Watson [6] can analyze the same volume of information in a quarter of the time while also diagnosing with greater precision. Watson is capable of ingesting more than 600,000 pieces of medical evidence, searching up to 1.5 million patient records, and reading 2 million pages from medical publications, which is a breadth of knowledge that no clinician can equal. Because of this, IBM Watson [6] can correctly detect lung cancer 90% of the time, whereas human physicians only achieve an accuracy rate of 50%. The other thing at which AI is best is creating variations, as after artificial intelligence has identified a pattern, it is able to use that pattern to instantly build many variants. An algorithm was able to pull from a database of patterns and colors to construct seven million various versions of the packaging for Nutella's product as part of a project that was given the name "Nutella Unica" [1].

Let's talk about what is fundamental to AI

When we talk about AI, we frequently refer to the algorithm that serves as its foundation. It might be beneficial to take a quick break and learn the definition of the word "algorithm" before we dive into all of that. Despite being used frequently, the expression doesn't always accurately convey much information about the issue at hand. This is partly because the word is rather vague in and of itself [7]. The following is what it is officially defined as: formula (noun): a method that involves following steps to solve a problem or achieve a goal, typically using a computer [8]. That perfectly captures everything. An algorithm is just a sequence of logical steps that describe how to carry out a specific task from beginning to end. This broad interpretation may be used to classify a cake recipe as an algorithm. It also counts if you have a list of directions you could give to a lost stranger. Theoretically, an algorithm can be thought of as any self-contained set of instructions for carrying out a specific task that has been precisely described. Examples of such lists include self-help books, IKEA manuals, and troubleshooting videos from YouTube. It is not exactly how the phrase is used in practice, though. When people discuss algorithms, they typically mean something a little bit more precise. Even though these algorithms are almost often mathematical objects, they nonetheless consist of a series of steps that must be taken sequentially, one after the other. They turn a set of mathematical processes into computer code, including equations, arithmetic, algebra, calculus, logic, and probability. Taking the order of operations is how this is accomplished. They receive facts from the real world, are given a goal, and are then set to work making the calculations required to achieve the goal. They are what establish computer science as a genuine academic discipline, and as a result, they have been the driving force behind many of the most astounding modern technological achievements. It would be difficult to tally all the different algorithms because there are so many of them. No one can agree on the best approach to group them all together because each one has different goals, eccentricities, advantages, and disadvantages. However, it can be useful to think of the actual jobs that they carry out in terms of the following categories [9].

Google Search [10] ranks your search results based on how closely they resemble the page you're looking for compared to alternative results. Based on your viewing history, Netflix [11] will suggest movies to you. All employ a mathematical technique to rank the inconceivably enormous number of options. Deep Blue also served as a sort of prioritization process, examining every move that could be made on the chessboard and deciding which one would have the most chance of victory [19]. Algorithms frequently require the removal of specific bits of information to concentrate on what is important and separate the signal from the noise. Sometimes they mean it in the most literal sense possible: speech recognition algorithms, such as those used by Siri, Alexa, and Cortana, must first separate your voice from background noise to begin deciphering what you are saying. Sometimes they do it metaphorically

[12]: Facebook and Twitter filter news depending on how closely they relate to your current interests to create your own personalized feed [13]. Algorithms exist that can read your handwriting and classify each mark on the page as an alphabetic letter. You can have algorithms label your vacation images for you. Algorithms exist that can read your handwriting and classify each meeting mark on the page [14]. Unsuitable content can be automatically removed from YouTube by algorithms.

The fundamental idea behind the concept of association is discovering and naming the relationships between diverse elements. The foundation of dating algorithms, including those employed by OKCupid, is association; these algorithms look for links between users and recommend matches based on those associations depending on the results of the inquiry [15]. A similar idea is used by Amazon's recommendation engine, which links your interests to those of prior customers. Because of the circumstances leading up to it, an intriguing buying proposal was made to you [16]. Most algorithms will be designed to perform a few of the functions in combination. Take UberPool, a ride-sharing service that connects potential passengers with other users traveling in the same route. It must examine all feasible routes to get you home once it has your starting point and ending point, search for connections with other users traveling in the same direction, and select one group to put you in. To make the travel as efficient as possible, all of this must be done while giving preference to routes with the fewest turns for the driver [17]. Some algorithms can therefore accomplish this how do they go about achieving it, then? Although there are essentially infinite options, there is a way to reduce the number of choices. The methods used by algorithms could be thought of as roughly falling under two primary paradigms. The naive method is one of these paradigms, and the greedy approach is the other.

Perspective two—AI cannot replace human

The modern era's inventions are comparable. A Nigerian named Chukwuemeka Afigbo created an automatic hand-soap dispenser that functioned wonderfully every time his white acquaintance ran their hand under the device; however, the device would not accept his hand because of his darker skin tone [19]. When Mark Zuckerberg was writing the code for Facebook in his Harvard dorm room in 2004, he could never have imagined that one day his platform would be charged with aiding in the sway of votes in international elections [20]. Each of these breakthroughs has an algorithm at its core. Algorithms are the intangible bits of code that make up the wheels and cogs of the contemporary machine era. Algorithms are as much a part of our modern infrastructure as ever before, having given the world everything from social media feeds to search engines, satellite navigation to music recommendation systems. They can be found in our residences, places of employment, and public buildings like courts and hospitals. Along with supermarkets and movie studios, law enforcement organizations also make use of them. They are aware of our likes and, using what they know about us, recommend films, books, and potential dates for us. They have a secret ability that allows them to gradually and subtly change the rules that define what it means to be human in the interim. There are many algorithms on which we increasingly rely, possibly without even realizing it. The police will utilize algorithms to determine who should be arrested, and these algorithms will put us in a position where we must decide between protecting the safety of crime victims and making sure that the accused are dealt fairly. Algorithms that judge use to determine the punishments for convicted offenders, and they will ask us to decide how our justice system should be organized. It will discover algorithms used by doctors to disregard their own diagnosis, algorithms built into autonomous vehicles that demand we make moral decisions, algorithms weighing in on our emotional outbursts, and algorithms having the power to destabilize our democracies. These are some of the talks of global village we live in.

Not that we're saying there's something fundamentally wrong with algorithms. There are numerous reasons to have a positive mindset and be hopeful about the future. There are no things or algorithms that are intrinsically good or evil. How they are used is what matters. The dynamic that exists between humans and robots must first be understood to generate an opinion regarding an algorithm. Every single one has a deep connection to the people who created it and utilize it. This makes it clear that this work is fundamentally about individuals and their interpersonal relationships. It discusses who we are, where we're going, what matters to us, and how technology is changing all those things. It is about how we connect with the algorithms that are already in place, working in tandem with us, enhancing our skills, catching our errors, resolving our problems, and posing fresh challenges as we go. The issue at hand is whether a specific algorithm offers society a net benefit. Regarding the situations in which you should trust a computer more than you should your own judgment, as well as the situations in which you should resist the urge to cede control to the machine. It involves looking closely at the algorithms to determine where their limitations are as well as at us to determine where ours are, identifying what is good and bad, deciding what type of world we want to live in, and making distinctions between these things.

Perspective three—AI will change everyone to be a designer

The New York Times recently revealed that Carnegie Mellon University intends to establish a research center that would focus on the ethics of artificial intelligence. This information should put to rest any skepticism that AI is already a reality. Both the Harvard Business Review and CNBC have begun setting the groundwork for what it implies for management, and CNBC has begun assessing AI equities that show promise [21].

> I argued, albeit in a somewhat pessimistic manner, that design will remain immune to the effects of AI in the relatively near future since excellent design requires both creative and social intelligence. However, this favorable picture for the immediate future did not completely allay all my anxieties. My daughter began her studies at a university this year with the intention of earning a degree in interaction design. When I first started looking at how AI might have an impact on design, I couldn't help but think about what kind of guidance I would give to my daughter and other designers of the future to assist them not only survive in a world dominated by AI, but also thrive in it.—Rob Griling.

Rob Griling argues that the following is what people ought to anticipate and get ready for in the year 2025. There will be no one left who is not a designer. Majority of jobs in the design industry nowadays require both creative and social intelligence. Empathy, innovative problem-solving, negotiation, and the ability to persuade others are all necessary components of these skill sets. The first change that artificial intelligence will bring about is an increase in the number of non-designers who work to improve their creative and social intelligence abilities to increase their marketability.

The implication for designers is that more people than just those employed in conventional creative industries will be trained to use "design thinking" processes to finish their work. If it was ever the case that designers had a monopoly on being the most "creative" people in the room, that will no longer be the case. To preserve their competitive advantage, designers will increasingly need specialized knowledge and abilities that allow them to contribute to a variety of scenarios. You can imagine a design-thinking-trained teacher experimenting constantly in a classroom with different frameworks for interaction to improve learning. Alternatively, a designer or hospital administrator is entrusted with redesigning the inpatient experience to enhance its effectiveness, usability, and general health results. This trend is already starting to emerge, as seen by the Seattle mayor's office's creation of an innovation team with the goal of solving the city's most pressing issues and concerns. The team includes both designers and design strategists, and it upholds the belief that "human-centered design" is the ideal strategy [21].

The d.school at Stanford University has been developing the creative genius of designers who have not received conventional training for more than ten years. New programs are also being created, like the Integrated Design and Management program that MIT offers. To better prepare future doctors, even medical schools are starting to include design thinking into their curricula. This not only emphasizes the wider significance of design but also demonstrates a novel possibility for academics in a wide range of disciplines to integrate training in creative intelligence and human-centered design into their individual courses.

We are aware of the usage of algorithmic techniques by tools like Autodesk Dreamcatcher [22] to provide designers with a more abstract interface for the development process. These tools may generate hundreds of variations of a design when given sufficient high-level guidance, boundaries, goals, and a problem to solve. It is up to the designers to pick their favorites or keep remixing them till they reach a spectacular design. Depending on the field of design, the effects are very different. The parametric movement in architecture, often known as "Parametricism 2.0," is a prime example of the doors that can be opened by creativity that has been technologically enhanced. Its effects are already being investigated and debated in the gaming industry, where we are now designing large virtual cities and virtual worlds [23]. Consider the computer game No Man's Sky, which is set in a deterministic open universe with more than 18 quintillion (1.81019) distinct planets and was procedurally generated. Despite not being a commercial success, the video game "No Man's Sky" does point in the way that will eventually take over the production of virtual content. In the future, the designer's job will be to establish the objectives, limits, and guidelines before scrutinizing and perfecting the designs produced by artificial intelligence [24].

A significant portion of the current excitement and advancement in the field of artificial intelligence can be attributed to deep reinforcement learning, a relatively young technique that only began to gain popularity three to four years ago. However, generative design methods are not new and have been around for a while. Deep Q [25] is an artificial intelligence program created by Google's DeepMind that plays Atari games and develops over time through deep reinforcement learning, eventually mastering astonishing talents like finding hidden game loopholes [26]. The fact that Deep Q is an Atari gameplayer is how it got its moniker. The key breakthrough with DeepMind's Deep Q and its successor AlphaGo [27], a computer program that plays the board game Go [28], is that neither product has artificial intelligence (AI) that is knowledgeable or skilled in gameplay. In the realm of artificial intelligence, this is a significant advancement. Furthermore, it is not essential for anyone to codify the game's rules. The only kind of input is visual, and the only controls and overall objective are to get the greatest score possible. As a result of their competitive character, video games make the perfect learning environment for artificial intelligence.

What about the layout, though? The role of the curator is now relevant in this situation. Designers of the future will train their artificial intelligence tools to manage design issues by creating models based on their own tastes. For instance, artifact's years of experience working in the healthcare industry have given them a deep and comprehensive view on the major difficulties in digital health design required for changing patient behaviors [29]. This is required to alter patient behavior. We may picture a future in which we will have access to sufficient data to enter behavior goals and ask the AI system to create a framework for a solution that gets around anticipated problems like confirmation bias and the empathy gap. We anticipate that this will occur in the not-too-distant future.

Most designers will experience a significant boost in productivity as a direct result of the fact that AI-driven parametric design makes it possible for them to produce millions of different variations of a design rapidly and simply. We will, suddenly, be able to investigate a vast number of potential detours in a far less amount of time than we require now. Amateur designers will have an easier time producing work that is passable, if not extraordinary, because of increases in productivity and improvements in equipment. This could potentially put pricing pressure on services provided by professional designers. The superstars of the design business, however, are not expected to be impacted by these changes, even though there will be less hurdles to learning and perfecting the trade. During the 90 s, we noticed a parallel movement in the fields of print and graphic design. The introduction of desktop publishing software led to the gradual disappearance of products at the cheaper end of the market. On the other hand, it encouraged a more widespread appreciation for design among all people, which in turn increased both the demand for and the ability to distinguish the very best designers. Superstar designers and the firms who invest in them will continue to have the upper hand, which will drive up the value of design brands until AI reaches the point where it can wow us with wholly original concepts.

A pessimist would argue that when many people lose their jobs due to automation powered by artificial intelligence, they will seek solace in a virtual reality setting. The demand for virtual worlds, goods, and experiences will consequently rise. I sincerely hope that we can prevent this terrifying future, but as virtual, augmented, and mixed reality become more common, this field will become the next frontier of design possibilities. In addition to being specific to this new medium, issues like how we connect with one another in virtual reality and how we create and share shared experiences call for abilities like creativity and social intelligence that are challenging for AI to replace. How do we interact with one another, for instance, in a virtual environment? A new demand for the more established design fields, including architecture, interior design, product design, and fashion design, can also arise as we hastily create virtual worlds. Given that, virtual worlds can replicate real-world settings, this might be the case [21].

Consider some of Michael Hansmeyer's [30] incomprehensible forms as an illustration of what may be achieved when people and machines collaborate to do tasks that neither could complete independently. Despite having millions of facets, these forms are too complex for a single person to build, yet they have the potential to revolutionize architecture. There is an unmistakable draw to learning how to boost our creativity as people and across all sectors of effort, even if this is just one example. We might imagine a time where our personal AI assistants regularly assess our work and offer ideas and suggestions for areas of growth since they have a thorough understanding of our influences, heroes, and inspirations. This new development would be highly beneficial. Imagine a world where problem-solving robots help humans view a problem from several angles and from a variety of conceptual frameworks. Simulated users test the things we've planned at this point of the development process to evaluate how they will work in several scenarios and offer adjustments before anything is even generated. While A/B testing bots are constantly looking for methods to improve our design work's performance slightly.

People who work in design are not threatened by AI; on the contrary, it offers them several opportunities, especially those who are responsible for designing the interfaces that emerging AI systems will use to interact with us. How do we create those tools for the creation of AI? How should we approach creating the intelligent platforms and services of the future, in your opinion? How could these systems be created so that they help us improve our humanity, creativity, and connections to the outside world? Although this is a difficult task, it also offers a fascinating chance for us and the generations that follow us [21].

1.2 Basics of Machine Learning

Building blocks of machine learning for products designers who want to build

The question of whether designers should also be able to write code remains hotly debated. Most individuals, regardless of where they stand on this topic, agree that designers ought to understand programming, this allows them to better comprehend limitations and have compassion for developers. In addition, it frees up designers to come up with creative solutions that aren't always pixel perfect. The same logic applies to why designers need to understand machine learning.

Machine learning is defined as the "subject area that investigates how machines can learn without being specifically taught." Even though Arthur Samuel first used the phrase more than fifty years ago [32], the most interesting uses of machine learning have just appeared in recent years. Machine learning is responsible for such innovations as digital assistants, autonomous vehicles, and spam-free email.

Machine learning has improved by a factor of ten in the past decade, thanks to the development of more efficient algorithms, more powerful hardware, and larger datasets. Companies like Google, Amazon, and Apple have been at the forefront of machine learning for years, but it's only in recent years that they've started making some of their tools available to developers. The moment to study machine learning and use it into your goods is now.

The importance of machine learning for designers

Now more than ever, designers may consider how machine learning can be used to enhance their products because it is readily available to them. Designers and programmers should have open communication about what can be done, how to get ready, and the expected results. The sample applications shown below are meant to spark such discussions.

Transform interactions into unique memories

To better cater to specific users, product developers can employ machine learning to tailor their offerings. We can then use this information to enhance features like suggested reading, search results, notifications, and advertisements.

Find the oddballs

The detection of out-of-the-ordinary material is greatly aided by machine learning. Financial institutions use it to identify fraudulent transactions, email providers use it to identify spam, and social media platforms use it to identify troubling content like hate speech.

Innovate new methods of communication

In recent years, advancements in machine learning have allowed computers to acquire a limited ability to comprehend human speech (natural language processing) and visual data (computer vision). Siri can now understand the command, "Siri, set a reminder…," dog photo albums can be made in Google Photos, and Facebook can now provide a written description of an image to the visually handicapped.

Give some clarification

Furthermore, machine learning can be used to better comprehend user segmentation. With this knowledge, analytics may be examined in greater detail across distinct demographics. This is where new features can be tested with a small subset of users before being released to the public.

Make your content ready

Thanks to machine learning, we can anticipate a user's next move. With this information in hand, we can better anticipate the needs of our users. For instance, if we can foresee which videos a user would like to watch, we can preload them in advance.

Classes of ML systems

There is a wide variety of machine learning algorithms to pick from, depending on the task at hand and the quantity and quality of the available data. Let's go through them in brief.

Supervised learning

Predictions can be made from appropriately labeled data with supervised learning. Informational tags or outputs are added to a set of examples to create a set of labeled data. Images with corresponding hashtags, for instance, or a home's specifications (e.g., square footage, neighborhood) and asking price. Supervised learning allows us to divide the data into groups or identify a trend by fitting a line to the labeled data. This is the line that allows us to extrapolate to undiscovered data. We can use new photographs to guess the hashtags people will use, and we can use a home's attributes to guess what it will sell for [31]. Classification is the term used when the target outcome is a set of labels or scores. When the target variable is a numerical metric, we use the term regression to describe the ensuing forecasting process.

Unsupervised learning

When we don't know how to categorize our data or we're unsure of the significance of certain outputs (such as image's hashtags or a house's price), unsupervised learning can help. Instead, we might see patterns in information that hasn't been categorized. In the context of online shopping, this can mean highlighting complementary products or suggesting related purchases based on previous customers' behavior. We refer to the pattern as a cluster if it occurs in groups. We refer to the pattern as an association if it takes the form of a rule (e.g., if this, then that).

Reinforcement learning

Without a preexisting dataset, reinforcement learning cannot function. Instead, we design a self-learning agent to glean information through trial and error in a rewarded setting. An agent can learn to play Mario by, say, earning a positive reward for collecting coins and a negative reward for walking into a Goomba. The method of reinforcement learning has proven to be successful in teaching computers, and it is based on the way people learn. Games like Go and Dota have been successfully taught to computers through reinforcement.

1.3 UserXperience and AI

UX state in AI

Algorithms are the fuel that keeps new technologies advancing into the future

The Turing test [34] was developed way back in the middle of the previous century with the purpose of determining whether or not a machine can trick a person into believing that there is another human being on the other side of the screen. In the year 1950, Alan Turing came up with the idea to test people's ability to determine whether or not something is a machine. The idea that machines are on the verge of becoming perfect replacements for humans or these interfaces that we completely can't tell the difference from despite the fact that our day-to-day experience shows us completely otherwise as an example of Siri and Alexa, but in general failed dismally at performing tasks that were beyond narrow in scope [33].

What is AI? What is machine learning?

The field of artificial intelligence, which has a history spanning several decades, was previously primarily focused on a discipline known as knowledge engineering. The notion of emulating human cognition in software, enabling computers to reason akin to human experts, encountered significant obstacles in past decades and made limited progress until the advent of sufficient processing power to analyze vast quantities of data, extract patterns, and make decisions based on them.

How do algorithms work?

When considering the various emerging forms of interaction, such as those reliant on camera vision, computer vision, natural language processing, and speech recognition, it becomes evident that these technologies are predominantly rooted in narrow artificial intelligence (AI) [35]. Pattern recognition is the cognitive ability to discern and comprehend recurring patterns, thereby facilitating an understanding of the underlying processes or phenomena. Machine learning has significantly enhanced the machines' ability to comprehend various forms of human-generated content, such as handwriting, speech, photos, and doodles, which were previously beyond their understanding.

An algorithm bears resemblance to a computer program. The process involves the application of a series of logical principles in order to reach a definitive inference. In a more practical sense, the document provides guidance on sequentially executing a series of computations or mathematical operations. In the field of machine learning, these functions typically correspond to mathematical models that are optimized to represent patterns identified in the real world through extensive datasets. In order to enable the machine to accurately identify the presence of dogs in images, it is imperative to expose it to an extensive dataset comprising a substantial number of dog images. This exposure will facilitate the machine's ability to recognize and discern the distinctive visual patterns associated with dogs. Given the limited scope of these algorithms, it is imperative that the patterns they seek are confined to a highly specific domain. If the machine is trained to recognize dogs, it is likely to exhibit a tendency to identify dogs as well as various other objects. The aforementioned phenomenon serves to strengthen these patterns and subsequently identifies their presence in various contexts, thereby exhibiting a notable advantage within a highly specific and limited domain. When attempting to expand the scope of a domain, it becomes apparent that biases may arise. Consequently, it is crucial to exercise caution in the manner in which we provide data to these machines. It is important to bear in mind that the responses generated by these machines are contingent upon the data on which they have been trained [33].

When examining the outcomes of numerous algorithms, it is common for them to indicate their level of confidence in the obtained results. As designers, we occasionally choose to present this confidence to users. For instance, Netflix [36] provides users with a percentage indicating the likelihood of a particular show being suitable for their preferences, such as 85%. Users then make adjustments based on this information. As viewers, we employ this tool to comprehend the likelihood of these matches, and endeavor to incorporate the machine's confidence into our cognitive processes while engaging in television consumption. There has been a notable effort to enhance the performance of web results, particularly in terms of optimizing page loading speed. Furthermore, this is highly commendable. This aspect contributes significantly to an exceptional user experience.

Another aspect of user experience pertains to the expeditious delivery of answers to individuals. As an illustration, when utilizing Google's [37] search engine, conducting a series of queries regarding the weather in New York would yield a compilation of web pages containing information pertaining to the current weather conditions in that location. Over time, Google's search results have evolved to prominently display the New York weather at the topmost section of the search page. When utilizing the Google Chrome web browser, a notable feature is the predictive search functionality. This feature is designed to display search suggestions within the search box, even prior to initiating the search, regardless of the user's location, including New York. The objective is to reduce the duration required to obtain the solution. A prerequisite for this task is a high level of confidence in the response, as its value lies solely in its accuracy. Hastily arriving at an incorrect answer can result in greater harm than investing a modest amount of time. It is preferable to acknowledge one's lack of knowledge rather than offer an incorrect response. Considerable contemplation has been devoted to algorithms, their association with news and information, and their potential to exert influence or provide biased information that reinforces preexisting beliefs, without necessarily fostering enlightenment among individuals aligned with either the right or left ideological spectrum.

An exemplary illustration of this phenomenon can be observed in the manifestation of "featured snippets," which are prominently displayed in a distinct box positioned above search results. The following responses are expedient in nature and may possess varying degrees of accuracy. As an illustration, in the event that an individual conducts a search query for "how to prepare an omelet" and receives a concise excerpt consisting of two sentences, this signifies that Google has identified and extracted the two sentences from online sources that provide a response to the inquiry. When engaging in UX design, it is imperative to prioritize the design process with a user-centric approach, ensuring that the design is tailored to the needs and preferences of the intended user base. It is advisable to refrain from making any assumptions. Expressing uncertainty and acknowledging gaps in knowledge is an acceptable practice, underscoring the necessity for reform in the presentation of such information to users at the forefront.

Challenges in presenting errors to users

Artificial intelligence (AI) is currently being implemented in a wider range of contexts than ever before. The increasing integration and automation of AI in our daily lives raise concerns about the potential consequences if it were to experience a failure, given the growing reliance on this technology by a larger population. As designers, our objective is to comprehend and manage the peculiar responses in order to mitigate their potential negative impact or provide appropriate contextualization. As designers, it is imperative to develop a comprehensive understanding of the algorithm we are working with, delineate the application's boundaries to align with its strengths, and ensure accurate reporting of its outcomes.

In what manner can individuals be informed about the circumstances under which they ought to exercise trust or skepticism towards a given response?

One aspect involves discovering novel methods to prompt individuals to activate their cognitive faculties for critical thinking. The opportune moment to involve human judgment arises precisely when an algorithm encounters failure or uncertainty. The reliance on machines for performing tasks should not be constant. The establishment of a collaborative alliance between humans and machines is imperative. Given that machines and humans both have limitations in certain areas, it is imperative to explore strategies for fostering a collaborative relationship between the two rather than solely relying on machines. This is particularly important as humans frequently delegate cognitive tasks to machines. The utilization of what Daniel Kahneman [38] refers to as "System 1" thinking, as expounded in his work "Thinking Fast and Slow," involves the conceptualization of the human brain as comprising two distinct systems. The initial system can be characterized as a passive one, as it operates on an automatic basis. When someone enquires about your name, you instinctively respond without much contemplation. Similarly, when someone enquires about your well-being, your automatic response tends to be "good." Are we truly performing well? Approximately 90% to 50% of the time, it is possible that our performance is suboptimal. However, it can be argued that this process resembles the activation of system one thinking. There exists a significant level of reliance on machines, which may be excessive. This precarious situation arises from our desire to cultivate trust and faith in these machines, despite their frequent failures. Consequently, skepticism has emerged, necessitating substantial efforts to address the concerns raised, as exemplified by the case of self-driving cars that you previously mentioned.

   In a broader context, rather than endeavoring to emulate human behavior and achieve human-like persuasiveness, it may be more advantageous to leverage the inherent strengths of machines and allow individuals to capitalize on their own unique capabilities. Artificial intelligence acknowledges that machines perceive the world and process logic in distinct manners compared to human beings [39]. Narrow machine learning demonstrates remarkable efficacy in addressing laborious or repetitive tasks that demand meticulous attention to detail, are susceptible to errors, or are devoid of human enjoyment. It effectively extracts valuable insights from such tasks. How can we effectively address the issue of individuals' lack of proficiency in certain job roles? The task at hand will necessitate a substantial amount of exertion. The task can be delegated to automated systems. In essence, it is imperative that these entities do not serve as a substitute for the human species, but rather function as a form of companion species. What are some effective strategies for collaborative engagement? A prevalent topic of discussion in the realm of artificial intelligence and machine learning pertains to the potential displacement of human labor. However, a more captivating line of inquiry revolves around the exploration of methods to enhance and augment human capabilities.

How can we design systems that enhance individuals' strengths and capabilities to a greater extent?

The systems that possess proficiency in determining normalcy exhibit the ability to identify patterns and establish a range of normalcy for a given phenomenon. Furthermore, these systems are capable of predicting the subsequent occurrence of normal events or comprehending deviations from the norm. These methods aim to identify recurring patterns in the manner in which tasks are executed. In the foreseeable future, machines are expected to excel in providing support for generating frameworks in the context of low-level production tasks. Airbnb [40] conducted notable experiments involving the utilization of machine models. In these experiments, sketches representing various elements from their design system were provided to the machine model. These sketches served as visual symbols for each element within the design system. Consequently, the system was able to interpret these symbols and extract the corresponding design patterns from the code. As a result, a web page could be generated from a simple whiteboard drawing, demonstrating the system's ability to efficiently construct web pages. The aforementioned web page may not be deemed flawless, yet it serves as a preliminary foundation. The utilization of this approach obviates the necessity of possessing a high-fidelity wireframe sketch, as it enables the designer to transition directly to the browser. This facilitates a direct and immediate dialogue between the designer and developer regarding the subsequent steps to be taken, thereby expediting the production process. There exists a noteworthy service known as "man," which bears resemblance to Uizard [41]. The process involves generating sketch files by associating symbols with user interface (UI) elements. There is a growing utilization of artificial intelligence in the domain of image cropping, wherein the technology is employed to determine optimal cropping strategies. Rather than relying on simplistic approaches such as center or corner cropping, AI systems possess the capability to comprehend the salient features of an image and crop accordingly. Consequently, tasks that were traditionally performed by humans in lower-level production roles are now effectively executed by machines. Tasks that require a significant amount of time, involve repetitive actions, demand meticulous attention to detail, are prone to errors, and lack enjoyment can be effectively assisted by machines.

Is it likely that digital designers who do not engage in the design of AI and machine learning will eventually face professional obsolescence?

The accessibility of machine learning is widespread, and it is crucial to consider the appropriate utilization of this technology. Consequently, it is imperative to discuss the involvement of designers and researchers in comprehending the suitable domains for the implementation of machine learning. The forthcoming generation of technology is being fueled by the oxygen of innovation, which is reminiscent of the significant impact that mobile technology had on our industry over the past decade. It is imperative to commence the process of delineating the distinct roles of designers within the emerging domain of artificial intelligence and machine learning.

What are the strategies for progressing further?

It is imperative to ascertain the strengths and limitations of these systems. Understanding the fundamental concepts of machine learning, including its definition, operational mechanisms, and the various classifications of models, is of considerable value. Designers may not necessarily require a comprehensive understanding of the mathematical principles underlying their work; however, it is advantageous for them to possess a comprehension of the strengths and limitations associated with these principles. For individuals working as visual user experience (UX) designers for web platforms, possessing a comprehensive understanding of the inherent characteristics of the web is highly advantageous. This includes recognizing its strengths and limitations in terms of interface rendering, as well as discerning which elements are challenging to present effectively and which are conducive to interface design. It is also advantageous to gain familiarity with various iterations of machine learning. Gaining practical experience with algorithms that an organization is actively engaged with and beginning to explore can be advantageous. It is imperative to comprehend the nature of the outcomes provided, the manner in which confidence is reported, and the accuracy of the obtained result.

User experience (UX) professionals' strategies for effective involvement in AI process

Considering the investigation of comprehending the individuals and procedures that we intend to assist, as well as identifying the areas lacking in understanding where machines could contribute by alleviating mundane tasks and providing valuable insights, are crucial aspects to contemplate. The field of user experience (UX), specifically its research aspect, plays a crucial role in identifying the areas where data scientists should direct their algorithms. Additionally, it contributes to the effective presentation of algorithmic outcomes, highlighting the capabilities of these algorithms in relation to how our interfaces typically present their recommendations. In essence, machine-generated interfaces exhibit a prevalent issue of excessive confidence, which does not originate from the algorithm per se. Rather, the interface portrays the information as unequivocally true, despite the algorithm's confidence level potentially being as low as 50%.

One of the tasks that must be undertaken by our industry is the initiation of the development of interface and interaction design patterns that effectively convey an appropriate level of confidence. Addressing bias in data is an area where user experience (UX) can provide valuable assistance. UX can contribute by mitigating the occurrence of unfavorable bias and by effectively highlighting instances of bias and the associated algorithmic processes when required.

Final words

It is not necessary for us designers to grasp the arithmetic that underpins artificial intelligence and other branches like machine learning and deep learning related to it, but it is helpful to understand what they are excellent at and what they are not good at. In the next chapter, we will be looking at the potential of AI for designers. Looking forward to meeting you all.

References

1. Jasmine, Oh. 2019. *Yes, AI will replace designers.* https://medium.com/microsoft-design/yes-ai-will-replace-designers-9d90c6e34502.
2. Janelle, Shane. 2023. *The AI weirdness hack.* https://www.aiweirdness.com/.
3. Mario, Klingemann. 2016. *Quasimondo.* https://underdestruction.com/category/research/.
4. Timm, Biggs. 2015. *Gorillas: Google photos uses racist tag on black friends, provoking backlash.* https://www.smh.com.au/technology/gorillas-google-photos-uses-racist-tag-on-black-friends-provoking-backlash-20150702-gi31y6.html.
5. Robert, A. Kornfield. 2011. *The new standard of care in medicine.* https://www.huffpost.com/entry/new-standard-medical-care_b_1117409.
6. IBM, Harnessing AI. https://www.ibm.com/in-en/watson.
7. Hannah, Fry. 2018. *Hello world: Being human in the age of algorithms.* 1st ed. W. W. Norton & Company.
8. Merriam Webster. https://www.merriam-webster.com/dictionary/artificial%20intelligence#:~:text=Definition%20of%20artificial%20intelligence,to%20imitate%20intelligent%20human%20behavior.
9. Diakopoulos, Nicholas. 2014. *Algorithmic accountability reporting: On the investigation of black boxes.* New York: Tow Center for Digital Journalism, Columbia University.
10. Google AI. *In-depth guide to how google search works, Google search central, documentation, Google developers.*
11. David, Giesbrecht. 2017. *This is how Netflix's top secret recommendation system works.* https://www.wired.co.uk/article/how-do-netflixs-algorithms-work-machine-learning-helps-to-predict-what-viewers-will-like.
12. Mona, Bushnell. 2023. *AI Faceoff: Siri versus Cortana versus Google assistant versus Alexa.* https://www.businessnewsdaily.com/10315-siri-cortana-google-assistant-amazon-alexa-face-off.html.
13. Lindsay, Kolowich Cox. 2021. *5 social media algorithms marketers need to know about in 2022.* https://blog.hubspot.com/marketing/how-algorithm-works-facebook-twitter-instagram.
14. Yuliia, Kniazieva. 2022. What is handwritten text recognition with OCR. https://labelyourdata.com/articles/ai-handwriting-recognition.

15. OKCupid. 2022. *How does OkCupid work? Our complete guide to match questions, the algorithm and setting up your account.* https://help.okcupid.com/hc/en-us/articles/688494321 3965-How-Does-OkCupid-Work-Our-Complete-Guide-to-Match-Questions-the-Algorithm-and-Setting-Up-Your-Account.
16. Kerbobotat. 2013. *Went to buy a baseball bat on Amazon, they have some interesting suggestions for accessories.* https://www.reddit.com/r/funny/comments/1nb16l/went_to_buy_a_baseball_bat_on_amazon_they_have/.
17. Sarah, Perez. 2017. *Uber debuts a "smarter" UberPool in Manhattan, TechCrunch.* https://techcrunch.com/2017/05/22/uberdebuts-a-smarter-uberpool-in-manhattan/.
18. Kasparov, Garry, and Deep Thinking. 2017. *Where machine intelligence ends and human creativity begins.* London: Hodder & Stoughton.
19. TheGoodKnight. 2012. *Deep blue versus Garry Kasparov game 2 (1997 Match).* https://www.youtube.com/watch?v=3Bd1Q2rOmok&t=2290s.
20. Karolina, Luzniak. 2022. *AI in product development—Examples and benefits.* https://neoteric.eu/blog/ai-in-product-development-examples-and-benefits/.
21. Rob, Girling. *AI and the future of design: What will the designer of 2025 look like?* https://www.artefactgroup.com/ideas/ai_design_2025/.
22. Rob, Girling. 2016. *AI and the future of design: What skills do we need to compete against the machines?* https://www.oreilly.com/radar/ai-and-the-future-of-design-what-skills-do-we-need-to-compete-against-the-machines/.
23. Patrik, Schumacher. 2016. *Parametricism 2.0—Gearing up to Impact the global built environment* (patrikschumacher.com).
24. VR game, No mans sky. https://www.nomanssky.com/.
25. Two minute papers. *Google DeepMind's deep Q-learning playing Atari breakout!—*YouTube.
26. Liat, Clark. 2015. DeepMind's AI is an Atari gaming pro now|WIRED UK. https://www.wired.co.uk/article/google-deepmind-atari.
27. Google DeepMind. https://www.deepmind.com/research/highlighted-research/alphago.
28. Go (game)—Wikipedia. https://en.wikipedia.org/wiki/Go_(game).
29. Emma, Tucker. 2017. *Artefact's aim health clinic can drive itself to a patient in need.* https://www.dezeen.com/2017/07/08/artefact-aim-health-clinic-drive-itself-to-patient-design-technology/.
30. Michael Hansmeyer. Building unimaginable shapes, TED Talk. https://www.ted.com/talks/michael_hansmeyer_building_unimaginable_shapes.
31. Drozdov, S. *An intro to machine learning for designers.* https://uxdesign.cc/an-intro-to-machine-learning-for-designers-5c74ba100257.
32. Samuel, A.L. 1959. Machine learning. *The Technology Review* 62 (1): 42–45.
33. Oppy, Graham, and David, Dowe. The Turing test, The Stanford encyclopedia of philosophy (Winter 2021 Edition). In Edward N. Zalta ed. https://plato.stanford.edu/archives/win2021/entries/turing-test/.
34. Narrow AI. *What is narrow AI?* https://deepai.org/machine-learning-glossary-and-terms/narrow-ai.
35. Netflix Research. *Learning how to entertain the world.* https://research.netflix.com/research-area/machine-learning.
36. Sejuti, Das. 2020. *From weather forecasting to 'Nowcasting': How Google is using ML to predict precipitation.* https://analyticsindiamag.com/from-weather-forecasting-to-nowcasting-how-google-is-using-ml-to-predict-precipitation/.
37. Kahneman, Daniel. 2011. 1934—Author. *Thinking, fast and slow.* New York: Farrar, Straus and Giroux.
38. Anirudh, V.K. 2022. *What are the types of artificial intelligence: Narrow, general, and super AI explained.* https://www.spiceworks.com/tech/artificial-intelligence/articles/types-of-ai/.
39. Benjamin, Wilkins. *Sketching interfaces generating code from low fidelity wireframes.* https://airbnb.design/sketching-interfaces/.
40. Tbeltramelli. 2017. *Teaching machines to understand user interfaces.* https://hackernoon.com/teaching-machines-to-understand-user-interfaces-5a0cdeb4d579.

41. Sarah, Karp. 2022. *A design manager reviews Uizard*. https://bootcamp.uxdesign.cc/a-design-manager-reviews-uizard-12e09adafe89.

Chapter 2
AI for Designers

You Designers are Afraid of ML. Are We?

2.1 Artificial Intelligence Meets Designers

Who has a fear of machine learning (ML)? To comprehend this technology, designers must delve slightly into the weeds. There are three essentials. There are three algorithms for machine learning: supervised learning, unsupervised learning, and reinforcement learning. Let's begin by discussing each one in detail.

Supervised learning

Supervised learning requires a comprehensive set of labeled data (labeled data refer to data that have been tagged, thereby categorizing the data. Consider spreadsheets of data). These tagged data are examined by algorithms, which then learn from data patterns and generate predictions. It is the most widespread form of machine learning; thus, it merits serious examination. For problems involving classification and regression, researchers would employ supervised learning. Classification is used whenever we want algorithms to anticipate the discrete category or "class" that new data will fall into. Regression would be used to forecast algorithmic outputs connected to real-valued numbers.

Let us start with a classification-based supervised learning technique. Imagine if your area becomes infested with snakes. Neighbors begin uploading images online to ascertain whether the snake in their yard is a copperhead, which is venomous, or a corn snake, which is innocuous. Their children play outside, and they must be informed. You decide to create an application to fix this issue. To construct the application, you need a method for distinguishing between snake species. First, you collect training data, which consist of a set of snakes individually identified as copperheads or corn snakes. This information represents our "ground truth." These designations can also be referred to as classes. Next, we must identify characteristics that the system can utilize to differentiate across classes. In this instance, we will determine two characteristics: length and mass.

© The Author(s), under exclusive license to Springer Nature Singapore Pte Ltd. 2024
M. H. Akhtar and J. Ramkumar, *AI for Designers*,
https://doi.org/10.1007/978-981-99-6897-8_2

A data scientist, developer, or designer may identify these features, or a team of domain-specific professionals could choose the features and label the data. The term for this is annotation. In this case, we may assemble a team of specialized herpetologists. The herpetologists would prepare a paper including a set of annotation guidelines. This procedure can be simple, but it can also be quite controversial. Participants may dispute about which characteristics are to employ or which dataset is best suitable. Experts in specific fields may also be concerned about sharing proprietary data information with the data science team executing the ML algorithms. Assume that our process goes without a hitch. After selecting the features, we purge the resultant labeled data of any errors, inconsistencies, missing data, or duplicates. We then randomize the data order and examine the training examples for bias. If there are an excessive number of copperheads in the training data, the system will be biased towards identifying more copperheads. This has occurred in various real-world situations, such as when an ML system was trained with too many samples of one skin tone and not enough of others [1]. We then set away a subset of the data for future system evaluation. Now comes the fun part. We select a learning algorithm and tell it to construct a model from the training set. In this case, we must instruct the system to divide the data into two distinct groups of snake species; therefore, it makes logical to employ a classification model. Note that, like with all ML techniques, this method generates a statistical model. We cannot attain 100% accuracy with any model, but we will select and adjust our model, so that it is as precise as feasible. ML, like design, is iterative. Throughout the process, we will need to modify the model to boost its predictive ability.

A classification model could utilize different strategies, such as decision tree, K-Nearest Neighbor, and Naive Bayes, to achieve our objective. Let's select a decision tree technique in this case. When this selected method is executed, the learning algorithms will develop a model that, in essence, draws a line over the data to produce a boundary separation. Depending on their characteristics, some of the snakes will be classified as copperheads and others as corn snakes. The algorithms will then compare the findings to the training data for accuracy and continually redraw the line to determine the ideal border separation; the place at which the greatest number of snakes are correctly categorized as belonging to their respective snake species. Humans oversee this procedure and attempt to enhance the outcomes. As stated previously, training data are important. To improve precision, we may need to add or remove training samples, particularly outliers. Or we may require feature adjustments. We might also use an alternative categorization technique, such as K-Nearest Neighbor or Naive Bayes. Each technique will have unique benefits and costs.

When we have confidence in the results, we may apply the model to the labeled dataset aside for the evaluation phase. Depending on the outcomes, we can either continue to build on the model or accept the accuracy rate. Eventually, we should be able to run any copperhead or corn snake through this system with a high rate of accuracy, not 100%, but a high percentage of accuracy. The computer's determination that a certain snake is a copperhead or corn snake is referred to as a forecast. Let us take a step back and admire the elegance of the resulting system. We no longer require human judgment to identify snake species. Moreover, there is no need to program manual rules. In addition, if the system has built-in feedback, the algorithms can learn from each misclassification and continuously improve their performance. The children and corn snakes are safe! In this case, two classifications (copperhead and corn snake) and two characteristics (mass and length) were employed. Consider that predictive algorithms can utilize numerous classes and hundreds or thousands of features.

Regression

Regression varies from classification in that it allows us to explore values between and beyond discrete classes and so generate predictions. This is possible since the output of regression is numerical (or continuous). We would not choose a regression model to categorize snakes since, in our case, we want each snake to be identified as belonging to one of two species. When we want the system to identify values between and beyond those initially set in the training data, we would utilize regression. Such values may include the future price of an item, the level of customer contentment, or a student's grades. In each of these situations, the algorithms would generate predictions based on the relationship between the continuous number and some other variable(s). We employed a classification approach to classify our input data (snakes) into discrete classes. Using regression, we want the algorithms to predict a particular numeric value as opposed to a class.

Common regression models include linear models, polynomial regression models, and neural networks for more sophisticated regression problems. For instance, regression might be used to predict the future price of a home based on the pace of local job development, the degree of customer happiness based on wait time, or the grade of a student based on the number of hours spent working in a studio. Consider the procedure for utilizing a model to forecast a student's grade (a numerical value between 0 and 100). First, we would match training samples of student grades, the dependent variable, with an independent variable, such as weekly studio time spent physically. After labeling some samples, we were able to train machine learning (ML) algorithms to predict the future grade of any student based on the independent variable studio time. Now, consider the big picture. As is the case with classification, predictive algorithms can solve considerably more complicated problems than our basic example. Instead of a single variable, in this case studio time, there could be multiple variables. Note that using regression, even though the algorithms can generate value predictions beyond those expressly expressed in the training data (i.e., we do not need to supply instances of every potential grade/hour combination), they do so based on labeled data and defined variables. This will change once unsupervised learning is included.

Supervised Learning—Key Takeaways

- Requires labeled training data.
- Has a well-defined objective: "Computer, look for these specific patterns in the data in order to predict x."
- The most prevalent type of ML at now.

Unsupervised Learning

An expert does not label the training data or supply the features/variables in unsupervised learning. Instead, the algorithms scan through the supplied data in search of unspecified regularities or patterns. So, for instance, instead of feeding our algorithms many images of snakes already tagged as corn snake or copperhead, we may feed the system many photographs of both types of snakes and ask the algorithms to look for distinguishing patterns. The algorithms would then select variables to uncover patterns. We wouldn't necessarily understand how the computers choose the variables, which could number in the hundreds. Even without a specific objective in mind, we may ask the algorithms to discover intriguing patterns. A supervisor is required for supervised learning; someone must label the data. Unsupervised learning is ineffective. Bypassing the supervisor can lead to startling and unexpected results.

When could humans engage in unsupervised learning?

In circumstances where the outcome is uncertain or the analysis process is too complex for a human to discover crucial differentiating labels and variables, we may choose unsupervised learning. Identifying a human face, detecting fraud, or anticipating consumer behavior is examples of such difficult circumstances. Unsupervised learning also plays a crucial part in deep learning, which will be covered further down. When machine learning algorithms find complicated patterns using thousands of variables, it can be difficult for humans to comprehend how the machines made their predictions. This is known as the Black Box Problem. If a targeted ad miscalculates our interests, this may not be a major issue. However, if we were denied parole based on a computer's forecast of recidivism, it would become a major issue, especially if no one could explain how the judgment was reached [2]. In addition to effectively assessing highly complicated scenarios, unsupervised learning algorithms can reduce the time and resources required by labeled data. Consider that for training, supervised learning algorithms require labeled data and identifiable features. Unsupervised learning algorithms can be fed cheaper, more easily obtained, unlabeled data. All the unstructured, multimodal data—images, sounds, and motion—mentioned previously become a valuable resource for these systems [3]. Three common unsupervised learning algorithms include clustering and dimensionality reduction. Let's investigate clustering.

Clustering

Using this technique, algorithms generate clusters or groups within the input data, i.e., data points with similar characteristics are grouped together and data points with dissimilar characteristics are separated. K-Means, Mean Shift, DBSCAN, Expectation–Maximization Clustering Using Gaussian Mixture Models, and Agglomerative Hierarchical Clustering are popular clustering algorithms. A retailer may employ clustering to segment its consumer base. They might input current client data points, such as age, zip code, gender, purchasing history, and then run clustering algorithms to create new categories. The algorithms could identify client segments that a designer or marketer would not often anticipate, or the model may identify outliers that can stimulate specialized markets [4]. Clustering could also be used to identify future trends, such as the professional turnover rate in each business [5]. Darktrace, a cybersecurity business, uses unsupervised learning to learn patterns of typical activity in a system and then searches for anomalies unusual behavior that could represent cyberattacks in the system. Instead, then relying on labeled data constructed from accumulated knowledge of past security concerns, the system can search for unidentified dangers [6]. Clustering can provide unexpected insights that beyond the average human viewpoint.

Unsupervised Learning—Key Takeaways

- Can find patterns in situations that are too complicated for human analysis.
- Does not require labeled data and/or defined features/variables.
- Can produce surprising insights.

Reinforced Learning

In both supervised and unsupervised learnings, algorithms use training data to create predictions. We can consider this information to be historical because it already exists in the world. Consider the following question: "Would you like to always be judged by your past actions?" Reinforcement learning algorithms, in contrast, do not make predictions based on historical data. Instead, these algorithms construct a prediction model on the fly through trial-and-error interaction with an environment. Here is how it operates: Its agent attempts various actions inside an environment. Each interaction with the environment generates comprehension, which serves as the input data. The agent then employs this information to iteratively change their activities to attain a certain objective as a reward.

Consider a video game as a simple metaphor for reinforcement learning. In this type of game, players traverse the surroundings by "playing" or engaging inside that environment, they discover which acts lead to victory and which do not. They repeat effective acts and use the resultant understanding to shape future behavior. This version of ML most closely resembles how humans learn knowledge. To learn how to roller skate, we do skates and give it a shot. Every time we fall, and we learn something about how to maintain our balance. Implementing this new information enables us to fulfill our objective, which is to go around on skates while keeping balance.

In 2017, DeepMind used reinforcement learning in conjunction with deep convolutional networks to defeat a human Go champion using the program AlphaGo. Later that same year, they launched AlphaZero, a more advanced program that excelled at chess, shogi, and Go. According to DeepMind, the training duration for each game was decided by its style and difficulty: nine hours for chess, twelve hours for shogi, and thirteen days for Go [7]. Compare this to the years of hard training required for human mastery. Imagine this technology being used to larger, high-impact situations requiring strategic action, such as global warming. Prior to this point in 2017, according to Pedro Domingos, author of The Master Algorithm, "supervised-learning individuals would make fun of reinforcement-learning individuals," DeepMind corrected everyone [8].

Since reinforcement learning algorithms may interact with an unpredictable digital or physical environment, they are commonly used in gaming AIs, logistics, resource management, robotics, and autonomous car navigation systems. Before releasing their products into the world, researchers can run these algorithms across millions of virtual simulations [9]. Mark Crowley, a computer scientist at the University of Waterloo, is now utilizing reinforcement learning to train virtual fires to forecast the route of future wildfires [10]. Uber refines its self-driving AI platform by subjecting it to thousands of virtual simulations and employing prediction algorithms to pit it against an equally sophisticated environment [11]. The training duration of these algorithms can be considerable, often longer than that of supervised learning systems, notwithstanding AlphaZero's fast speed—but the capacity to respond to surroundings in real time can outweigh this lengthy training period [12].

Common strategies for reinforcement learning include of Deep Deterministic Policy Gradient, Q-Learning, State–Action–Reward–State–Action, and Deep Q-Networks. Note that in each of these methods, like in unsupervised learning, the process is not supervised. Due of the absence of a supervisor, bias cannot be introduced into the system. To clarify, this does not eliminate all opportunities for prejudice, but it does eliminate one prevalent route [13]. This lack of a supervisor also means that reinforcement learning can introduce alien ways for "winning," i.e., obtaining the reward, and subsequently teach these strategies to people. In the preceding example of AlphaGo and AlphaZero, Go champions analyzed the winning techniques employed by these algorithms and subsequently adopted them in their own games. However, some of the strategies were ineffective because, despite their best efforts, human players were unable to comprehend them [14]. With less overt bias and the introduction of mind-boggling new tactics, reinforcement learning is currently attracting significant research funding [15].

Reinforcement Learning—Key Takeaways

- Model interacts sequentially with an environment over time.
- Eliminates supervisor bias.
- Can introduce alien strategies that prove beneficial to humans.

Conclusion

Although we examined each ML category separately, other researchers combine them in practice. Semi-supervised learning refers primarily to the combination of supervised and unsupervised learnings, but researchers can also combine all three types to achieve their objectives [16]. This process can be muddled and intricate, needing an in-depth understanding of mathematics and statistics as well as a healthy dose of intuition. Programming explicit logic-based instructions is a distinct skill set than training a set of algorithms. This moves from programming to training results in a less distinct and more difficult to regulate connection with robots. The site director of Wired, Jason Tanz [17], argues, "If in the past programmers were viewed as gods because they wrote the laws that govern computer systems, now they are viewed like parents or dog trainers. This is a lot more mysterious relationship, as any parent or dog owner will tell you." This sort of liminal region is ideal for designers. Together with data scientists, we may begin designing user experience and interface options that map out these emerging human–machine connections.

2.2 Designing a Machine Learning Model

Designing a ML model in a fun and interactive way

In this study, we aim to develop a machine learning model that can effectively differentiate between various residential properties using a given dataset for Delhi (indicated in green) and Mumbai (indicated in blue). The goal of machine learning[a] (a—machine learning concepts have emerged in various academic areas, including computer science, statistics, engineering, and psychology, resulting in the use of multiple terminologies) is to teach computers to autonomously recognize and exploit patterns in data using statistical methods of instruction. Using these methods, one can produce quite precise forecasts.

An intuition to classification

Let us assume that we have to decide whether a house is in Delhi (indicated in green) or Mumbai (indicated in blue). Classification is a type of machine learning activity that involves sorting data into categories. With the city's steep terrain, the height of a house could serve as a useful identifier. An argument might be made, using the elevation data, that any home in Mumbai with an elevation greater than 280 m should be considered one (Fig. 2.1).

Adding spices to the ML dish we are cooking

The addition of this new dimension provides greater depth. Mumbai apartments, for instance, are known for their high cost per square foot. A scatterplot of elevation and price per square foot allows us to easily identify residences at different elevations. According to the numbers, Mumbai is home to the highest price per square meter of any city in India, at $600. Features, predictors, and variables are all terms used to refer to the dimensions of a dataset.

Drawing the estimate boundaries

If you make a scatterplot (Fig. 2.2), you may use your observations of high elevation (>280 m) and high cost per square foot (>$600) to demarcate different areas. In the green and blue areas, you might imagine dwellings in Delhi (indicated in green) and Mumbai (indicated in blue). The core of statistical training is the application of mathematics to the task of locating limits in data. Obviously, we will need more data to tell the difference between homes of varying elevations and per-square-foot costs.

Fig. 2.1 Unsupervised learning

There are seven dimensions in the dataset that we are utilizing to build the model. The principle to develop a model is to train it. A scatterplot matrix is being utilized to visually represent the correlations between each pair of parameters (Fig. 2.3). The data exhibit discernible patterns, but its bounds remain hazy.

Bring the underdog to the arena

Machine learning is used to analyze data for patterns. Many modern approaches to machine learning rely on statistical learning to locate limits. An application of machine learning is the decision tree. Decision trees are a machine learning technique that is considered to be generally approachable, albeit rudimentary, as it analyzes variables individually.

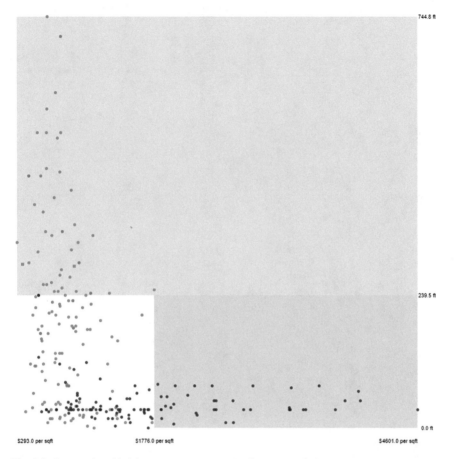

Fig. 2.2 Scatter plot of height versus cost per square foot

Finding better boundaries

To see whether we can refine our initial guess, let us look again at the 280-m-high limit that was proposed. Obviously, a new point of view is needed. For a clearer picture of how often houses appear at different heights, let's convert our representation into a histogram (Fig. 2.4). Although 280 m is the height of Mumbai (indicated in blue) City's tallest dwelling, most of the city's residences are much more modest in stature.

It is time to put the fork in the plate

One way to establish data patterns is with a "if–then" statement, which is what a decision tree employs (Fig. 2.5). As an illustration, if a house's elevation is more than a certain threshold, it is likely located in Delhi (indicated in green).

Forks are statements used in machine learning that divide the data in two different directions at the given value. The value at which the two paths diverge is known as the split point. Houses on the left of the dividing line are grouped together, while

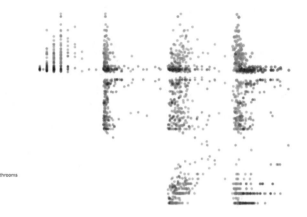

elevation

year built

bathrooms

bedrooms

price

square feet

price per sqft

Fig. 2.3 Scatter plot for different factors mapped

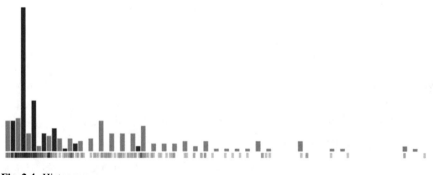

Fig. 2.4 Histogram

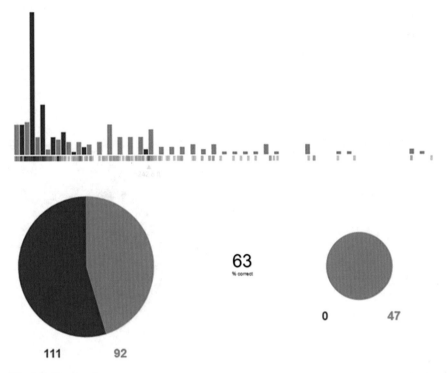

Fig. 2.5 Pie chart I

those to the right are grouped together. In a decision tree, a boundary is represented by a split point.

Trade-offs

Choosing a cutoff has consequences. Some homes in Delhi (indicated in green) were misidentified as being in Mumbai (indicated in blue) City because of our initial separation (280 m). The misclassified homes in Delhi are represented by the enormous chunk of green in the pie chart's leftmost section. The medical term for this is "false negative." A division that is designed to encompass all Delhi (indicated in green) residences will in fact include many Mumbai (indicated in blue) City dwellings. Mistakenly positive results are referred to as false positives (Fig. 2.6).

The best split

Each branch's final products should be as like one another as possible at the optimal split. You can choose from a variety of mathematical techniques for finding the optimal division[b] (b—for further exploration of the computation of the optimal split, it is recommended to conduct a search on the concepts of "gini index" or "cross entropy"). Even a perfect split on a single characteristic does not adequately

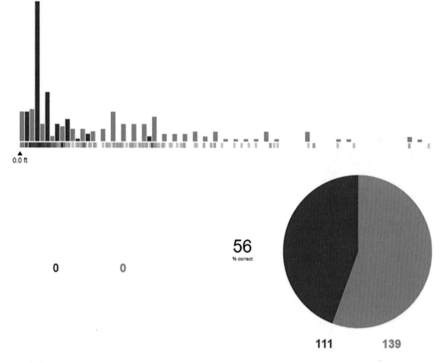

Fig. 2.6 Pie chart II

distinguish the homes in Delhi (indicated in green) and Mumbai (indicated in blue) (Fig. 2.7).

Recursion

The algorithm iteratively applies the steps taken to create the initial splits to each of the resulting data subsets to generate additional splitting criteria[c] (c—computers possess a notable proficiency in implementing statistical learning approaches due to their capacity to perform repetitive activities with remarkable speed and unwavering focus, so circumventing the limitations of human susceptibility to boredom). Recursion describes this pattern of repeating, which is a common feature of learning models. The optimal split will change depending on which tree node you examine[d] (d—the technique outlined in this paper can be classified as a greedy algorithm due to its utilization of a top-down strategy for data partitioning. Put differently, the objective is to identify the variable that maximizes the homogeneity of each subgroup at a given point in time).

One of the finest variables for the next if–then statement is the price per square foot, and at $600 per sqft, it is ideal for describing residences at lower elevations. The median price of a home at a high elevation is $180, 000 approx. (Fig. 2.8).

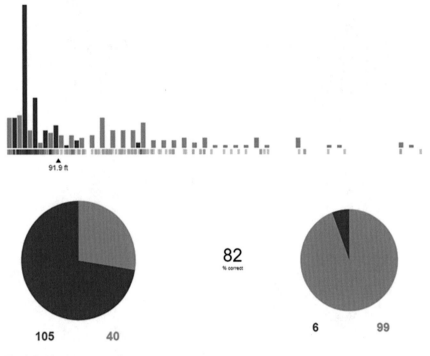

Fig. 2.7 Pie chart comparison

Growing a Tree

When a tree has more branches, it can make better predictions because of the more data it collects (Figs. 2.9, 2.10, 2.11, 2.12, 2.13 and 2.14). Accuracy rises to 84% when further splitting is performed on the tree (Fig. 2.9).

To get to 96%, we need to add numerous additional layers (Figs. 2.10, 2.11).

You could keep adding nodes (Fig. 2.12) to the tree until every forecast it makes is spot on (100%); for example, you could have every branch finish only in Delhi (indicated in green) or Mumbai.

These very last twigs on the tree are known as leaf nodes (Fig. 2.13). Each leaf node in our decision trees will be labeled with the predominant type of home.

Making predictions

Each data point is processed by the branches of the freshly trained decision tree model to establish whether a residence is in Delhi (indicated in green) or Mumbai (Fig. 2.14).

Here, the training data for the tree are shown as it passes through the tree. As its name implies, training data were utilized to teach the model (Fig. 2.15).

The tree was trained until it precisely mapped each training data point to its corresponding city. Applying the tree to data points it has never seen before is the

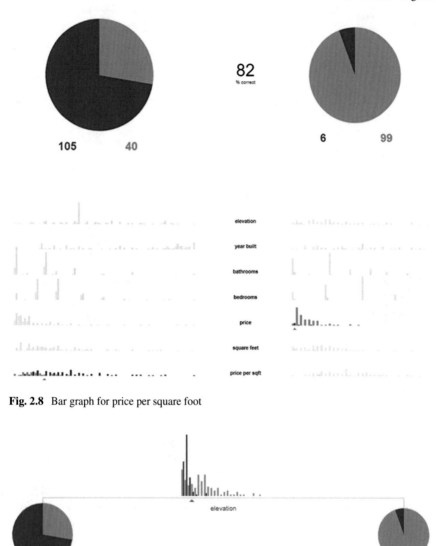

Fig. 2.8 Bar graph for price per square foot

Fig. 2.9 Tree top

only way to properly evaluate its ability to handle novel information. The term
"test data" is used to describe this information. The ideal tree would have the same
efficiency on both known and unknown information. So, this one is less than ideal
(Fig. 2.16).

Caused by overfitting, several mistakes have occurred. The model has been trained
to prioritize every piece of information, no matter how unimportant it ultimately
turned out to be, in the training data.

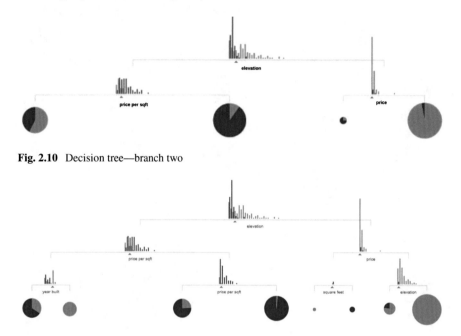

Fig. 2.10   Decision tree—branch two

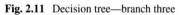

Fig. 2.11   Decision tree—branch three

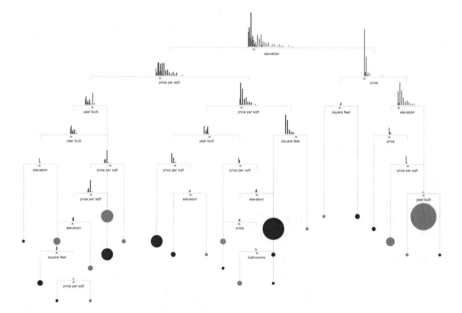

Fig. 2.12   Decision tree—branch eight

Fig. 2.13 Decision tree—last nodes

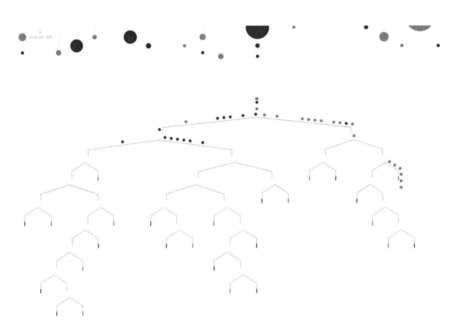

Fig. 2.14 Data point in branches

2.3   Being Humanist with AI

What makes people feel creeped out while using AI applications?

> Let machines do what they are really good at, and humans do what they are really good
> at—Author

Have you ever experienced this phenomenon? An enterprise application has been
developed, with its primary distinguishing feature being the integration of artificial
intelligence (AI). The diligent efforts invested in tailoring the software to individual
users based on available data are subsequently undermined as the agreement begins
to unravel. The experimental implementation of the aforementioned system among
the company's workforce has elicited a sense of apprehension and unease among
the employees. The Avatar has been described as unsettling, with an indescribable
quality that evokes discomfort in users. The seminal scholarly article in the field of

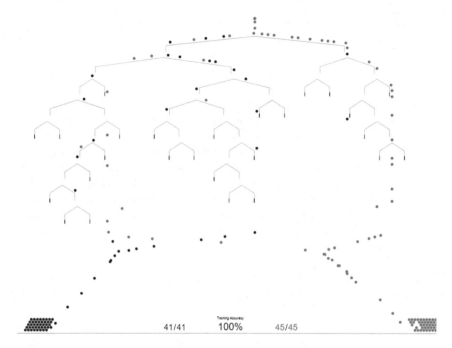

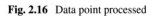

Fig. 2.15 Data point in processing stage

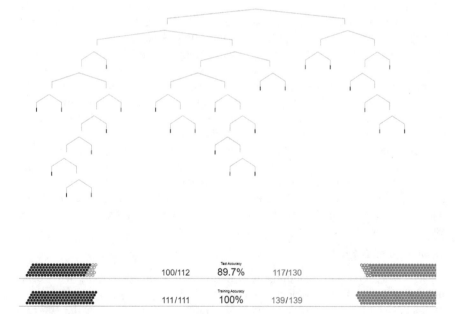

Fig. 2.16 Data point processed

psychology pertaining to the concept of creepiness is titled "On the nature of creepiness." This paper posits the existence of four fundamental factors that contribute to the perception of creepiness. The aforementioned individuals elicit feelings of fear or anxiety within us. It is imperative to ascertain whether there is a valid basis for such emotions in relation to the individual in question. The attribute of creepiness is commonly attributed to the inherent nature of the individual, rather than being solely attributed to their observable actions [18]. It is plausible to speculate that they harbor a sexual interest in our person. Research findings indicate that in surveys, men are often perceived as exhibiting more creepiness than women. Consequently, employing a female voice for virtual assistants such as Siri, Cortana, Alexa, and other AI voice systems has the potential to mitigate the perceived level of creepiness for female users of these products. Given the multitude of other contributing factors, it is logical that the presence of artificial intelligence (AI) has the potential to induce anxiety among individuals. This is a novel concept that individuals have yet to become acquainted with. The autonomous nature of the machine learning model raises concerns regarding user trust and their willingness to rely on its capabilities. There are various factors that contribute to the perception of creepiness. Hence, it is imperative to examine the intricacies of each factor from the standpoint of a designer.

Good Design—*Visual design*

When individuals experience a sense of unease towards an individual, characteristic attributes that may contribute to this perception include the presence of greasy hair, a distinctive smile, protruding eyes, elongated fingers, disheveled hair, notably fair complexion, under-eye bags, unconventional attire, and the donning of soiled garments. In the realm of computer technology, this serves as compelling evidence that a meticulously crafted homepage, landing page, or initial impression for your product is indispensable. According to empirical research, it has been observed that the human brain is capable of forming initial assessments regarding the trustworthiness of an individual's face in a remarkably short span of time, specifically within a mere 39 milliseconds [18]. Upon encountering an anomaly, our cognitive processes prompt us to initiate the deconstruction of facial features, subsequently leading to the deconstruction of the individual as a whole. The aforementioned statement holds true when employing AI software. If the software exhibits a high level of design quality, it is indicative of a positive initial impression. The significance of this phenomenon lies in its association with the "halo effect." In essence, individuals who possess physical attractiveness were perceived as being trustworthy, irrespective of their status as Nobel laureates or individuals with criminal backgrounds.

What is good design?

Let us gain a deeper understanding by illustrating with a counterexample. In the event that the application was developed and subsequently displays all data on the screen without due consideration for its visual presentation, it is plausible that the user may harbor suspicions regarding potential oversights resulting from hastiness during the development process. Consequently, concerns may arise regarding the vulnerability of one's data. In academic contexts, it is crucial to emphasize the significance of details. It is imperative to direct one's focus towards them. Ensuring the accuracy of various details during the initial interaction with the user can significantly enhance the quality of their initial impression.

Sound usage

The subsequent domain that warrants attention for effective design pertains to the utilization of language. Individuals were actively seeking indications of benevolence or hostility in the facial expressions of the individuals they were assessing. Similarly, this principle applies to software. The use of aggressive or abrupt language has the potential to evoke a sense of unease or tension in individuals. In addition to conventional software, artificial intelligence (AI) faces the challenge of overcoming its preconceived notions derived from cinematic portrayals. Consequently, it is imperative for AI to elucidate its processes in a manner that is both tactful and devoid of euphemistic language. One additional factor that elicits discomfort among individuals is the behavior of an individual who invades personal space by standing in close proximity to a friend or employs excessively affable language. An exemplary illustration can be observed in the behavior of a salesperson who exhibits an unsettling demeanor by promptly assuming a friendly disposition. In the event that your product incorporates any sales, it is imperative to ensure that the language employed within the product is appropriately refined and adheres to professional standards. This phenomenon contradicts the prevailing inclination towards employing overtly amicable language, which may contribute to the eerie nature often associated with AI.

When considering the subject of communication, it is important to also acknowledge the significance of non-verbal cues. It is advised to refrain from imitating non-verbal cues, such as hand gestures or body language, that do not convey any meaningful information. Although essential for facilitating seamless human interaction, the topic at hand is beginning to delve into the concept known as the uncanny valley, which will be discussed in detail shortly. It is worth noting that navigating this concept can be quite challenging. Even in the event of accurate execution, individuals may develop a sense of skepticism and experience a physical reaction characterized by a sensation of coldness [19]. An illustrative instance of this phenomenon can be observed in the case of Duplex, a speech system developed by Google, which was showcased at the 2018 IO conference. The speech engine's remarkable precision, encompassing pauses, sighs, and verbal fillers such as "uhms," effectively persuaded individuals lacking awareness of its non-human nature. However, according to the various critiques of the system, users expressed their discomfort with its behavior due to its tendency to emit audible sighs and utter the interjection "hmm." Technological advancements can surpass the comprehension of individuals, resulting in phenomena that appear as inexplicable sorcery, thereby instilling fear and apprehension among the populace.

One additional non-verbal cue that elicits discomfort among individuals is when an individual exhibits laughter at unexpected moments or demonstrates inappropriate emotional responses. When executed incorrectly, the action may be perceived as unsettling or disturbing. The incorporation of emotions into machine learning poses significant challenges, necessitating cautious progress and thorough validation of any modifications made. In order to maintain consistency in the emotional tone throughout the entire user journey, it is essential to conduct user tests after each addition and regression test for compatibility of each task. The final aspect that transitions from the realm of research to design pertains to the prevalence of eerie pastimes centered around the act of collecting various items. In light of this observation, it is highly probable that your product is amassing a substantial amount of data pertaining to these activities. The recommendation is to refrain from discussing the entirety of the information being gathered. The inclusion of a feature that emits an audible signal each time that the application acquires new information about the user is unnecessary.

Good interaction

There exist certain factors that contribute to the perception of individuals as being unsettling or eerie during interpersonal interactions. One notable cautionary observation arising from interpersonal interactions is the occurrence wherein an individual renders it exceedingly challenging to disengage from a conversation without conveying a sense of impoliteness. Additionally, another concerning behavior entails the persistent redirection of the discourse towards a singular subject matter. It is imperative to ensure that the product is equipped with a mechanism for engaging in human communication. There is a prevalent perception among a significant number of individuals that artificial intelligence (AI) is incapable of completely eliminating call centers from one's product. Even in the presence of a flawless machine learning model, user support will remain necessary. Similarly, it is advisable to avoid utilizing a conversational interface as the sole means of interaction. The lack of functionality in both office and public settings renders the imposition of user interaction in such environments a source of discomfort. Furthermore, it is imperative that our application refrains from repeatedly presenting the same script to the user until their consent is obtained. This particular aspect has the greatest impact on sales applications. In the event that the product is exclusively intended for sales purposes, it is imperative to bid farewell to potential customers who exit the sales funnel. In the event that the product possesses additional functionality, it is advisable to redirect the user back to their previous location within the interface, thereby preventing them from remaining within a sales funnel.

Another unsettling aspect involves soliciting personal information. The excessive number of inquiries poses a challenge. The issue lies in the nature of questions that are excessively personal. Consider the flashlight application on our mobile devices that is seeking permission to access our contact list. It is necessary to conduct a thorough examination of the inquiries posed to the user. Does this information align with existing knowledge from other sources? Can the inference of the answer be derived by combining the responses obtained from two previously posed questions? Furthermore, the final instance of social blunder pertains to excessive discussion of one's personal life. Startups are significantly impacted by this phenomenon. We express our admiration for your product and company. However, it is important to refrain from imposing your narrative onto your users. If individuals inquire or express interest, it is acceptable to provide personal information. However, it is advisable to confine personal details to a designated section or context.

Uncanny valley

The concept of the uncanny valley pertains to the notion that individuals exhibit positive responses towards humanoid figures and interactions, up until a certain threshold where these entities become excessively human-like. At this juncture, the subtle disparities between individuals of the human and non-human variety, such as an ungainly manner of walking, an incapacity to employ suitable eye contact or speech patterns, assume greater prominence due to the overall congruity of their other attributes, thereby engendering feelings of unease and disquietude. The concept was initially conceived by Masahiro Mori, a Japanese roboticist, in his essay published in 1970, wherein he predicted the difficulties that would be encountered by creators of robots. The underlying rationale for the utilization of cartoon avatars, cartoon-like robots, and the disconcerting nature of video game glitches can be attributed to this phenomenon. The aforementioned instance, as previously discussed, pertains to Google's duplex technology. The auditory output exhibited a high degree of resemblance to human speech, albeit not reaching a level of fidelity that could be classified as fully human, thus placing it firmly within the realm of the uncanny valley. Naturally, the means of circumventing this predicament entails refraining from entering the valley. Similar to many other products, users have the option to incorporate exaggerated levels of detail, utilize animal representations instead of human figures, or even forgo the use of avatars altogether. Additionally, it is crucial to inform the user about the constraints or restrictions associated with the subject matter. When users have high expectations for perfection, the uncanny valley phenomenon appears to be more pronounced compared to when they possess knowledge about the boundaries of realism [18].

User testing

Throughout our discussion, we have extensively addressed various factors that should be considered in order to eliminate any elements of creepiness from our product. How can one ascertain its efficacy? Undoubtedly, conducting user testing is imperative. In addition to various methodologies employed in user testing, it is worth noting that a quantitative approach exists for assessing user experience, specifically in relation to feelings of discomfort. Notably, individuals may experience a sensation of coldness when subjected to unsettling or creepy stimuli. In the event that alternative approaches prove ineffective, it is advisable to obtain a foundational understanding of our users' perceptions regarding the temperature. Next, assess the revised edition for potential enhancements [19].

Conclusion

In synthesis, it can be conceptualized that the sensation of creepiness serves as a cognitive mechanism employed by the human brain to discern potential threats or hazards. Our brain has a tendency to alert us to anything that deviates from the norm or presents unexpected elements. Humans have developed a tendency to prioritize the detection of potential threats in situations that lack clear information. In order to mitigate potential issues, it is recommended to employ a human-centered design approach. Much of the sense of unease arises from prioritizing our own needs over those of the customers. Therefore, it is imperative to ensure that our focus remains on addressing the customers' problems. Merely employing the machine learning model for the purpose of marketing products or amassing user data is likely to elicit cautionary concerns within the user's cognitive framework. Subsequently, it is imperative to delineate the trajectory by which the user anticipates resolving their predicament. Conduct a comprehensive evaluation of the user journey to ensure its effectiveness and efficiency. The purpose of this measure is to mitigate unforeseen deviations in the course of the journey.

Maintaining the user's awareness of their current position in the journey, alternative destinations, and available actions at any given point is considered a fundamental principle of effective design. If the standard software protocols are not adhered to, the resultant software may exhibit a lack of clarity and coherence. If the user possesses knowledge regarding the integration of machine learning within the software, they may perceive the artificial intelligence as being in control of the decision-making process. The absence of regulation results in a disconcerting sensation. In order to maintain user autonomy, it is imperative to allow individuals to determine the specific information that is collected. Inquire whether they desire enhanced levels of interaction facilitated by the process of data collection. Inquire as to whether the individual desires to receive customized advertisements. In the event that data collection is necessary for the proper functioning of our product, it is advisable to inform the customer of the product's inability to proceed rather than attempting to collect data without prior notification to the customer.

Final words

We designers need to know the application part of machine learning and make it visible in terms of histograms, bar graphs, histograms for technical team to understand and visual illustrations for the end user to have the best experience.

References

1. Joy Buolamwini. *Algorithmic justice league*. https://www.ajl.org/.
2. Cathy O'Neil. 2016. *Weapons of math destruction: How big data increases inequality and threatens democracy*, 272. New York: Crown Publishers. (ISBN 978-0553418811). https://doi.org/10.5860/crl.78.3.403.
3. Ethem Alpaydin. 2017. *Machine learning*. The New AI-The MIT Press.
4. Ibid, 112.
5. Chin-Yuan Fan, Pei-Shu Fan, Te-Yi Chan, and Shu-Hao Chang. 2012. Using hybrid data mining and machine learning clustering analysis to predict the turnover rate for technology professionals. *Expert Systems with Applications* 39 (10): 8844–51.
6. Steven Melendez. 2016. *This security company based its tech on the human immune system*. Wired. https://www.fastcompany.com/3062095/this-securitycompany-based-its-tech-on-the-human-immune-system.

7. David Silver, et al. 2018. *AlphaZero: Shedding new light on the grand games of chess, shogi and go*. DeepMind blog. https://deepmind.com/blog/alphazero-shedding-new-light-grand-games-chessshogi-and-go/.

8. Pedro Domingos. 2015. *The master algorithm: How the quest for the ultimate learning machine will remake our world*. New York: Basic Books; see also Karen Hao. 2019. *We analyzed 16,625 papers to figure out where AI is headed next*. MIT Technology Review. https://www.technolog yreview.com/s/612768/we-analyzed-16625-papers-to-figure-out-where-ai-is-headednext/.

9. Will Knight. 2019. *Facebook's new Poker-playing AI could wreck the online Poker industry, so it's not being released*. MIT Technology Review. https://www.technologyreview.com/s/613 943/facebooks-newpoker-playing-ai-could-wreck-the-online-poker-industryso-its-not-being; see also Tom Simonite. 2017. *AI sumo wrestlers could make future robots more nimble*. Wired. https://www.wired.com/story/ai-sumo-wrestlers-could-make-future-robots-morenimble.

10. Matt Simon. 2018. *How supercomputers can fix our wildfire problem*. Wired. https://www.wired.com/story/how-supercomputers-can-help-fix-ourwildfire-problem.

11. Sean Captain. 2018. Here's how to avoid more self-driving car deaths, says Uber's former AI chief. Fast company. https://www.fastcompany.com/40547165/ubers-former-head-of-aiheres-how-to-avoid-more-accidents.

12. Alpaydin. *Machine learning*, 136.

13. Ibid.

14. Dawn Chan. 2017. The AI that has nothing to learn from humans. Atlantic. https://www.theatl antic.com/technology/archive/2017/10/alphagozero-the-ai-that-taught-itself-go/543450/; see also David Silver, et al. 2018. A general reinforcement learning algorithm that masters chess, Shogi, and go through self-play. *Science* 362 (6419): 1140–44. https://science.sciencemag.org/content/362/6419/1140.abstract.

15. Karen Hao. 2018. The rare form of machine learning that can spot hackers who have already broken in, MIT TECH. REV. https://www.technologyreview.com/2018/11/16/139055/the-rare-form-of-machine-learning-that-can-spot-hackers-who-have-already-broken-in/.

16. Jordan, M.I., and T.M. Mitchell. 2017. Machine learning: Trends, perspectives, and prospects. *Science* 349 (6245): 258. https://science-sciencemag-org.prox.lib.ncsu.edu/content/349/624 5/255.

17. Jason Tanz. 2016. *Soon we won't program computers, we'll train them like dogs*. www.wired.com/2016/05/the-end-of-code.

18. Leander, N.P., T.L. Chartrand, and J.A. Bargh. 2012. You give me the chills: Embodied reactions to inappropriate amounts of behavioral mimicry. *Psychological Science* 23 (7): 772–779. https://doi.org/10.1177/0956797611434535.

19. Lauren Goode. 2018. *How Google's eerie robot phone calls hint at AI's future*. https://www.wired.com/story/google-duplex-phone-calls-ai-future/.

20. Zhong, C.B., and G.J. Leonardelli. 2008. Cold and lonely: Does social exclusion literally feel cold? *Psychological Science* 19 (9): 838–842. https://doi.org/10.1111/j.1467-9280.2008.021 65.x.

Chapter 3
AI in Product Design

Do Product Designers Use AI? What Do You Think?

3.1 AI Boosting Product Design

A computer would deserve to be called intelligent if it could deceive a human into believing
that it was human—Alan Turing

***Case study 1: Three-dimensional digital knitting of intelligent textile sensor for
activity recognition and biomechanical monitoring*** [1]

A methodology for the creation of a seamless and scalable intelligent textile based
on a piezo-resistive matrix was proposed. This approach involves the utilization of
both digital flat-bed and circular knitting machines. Through the integration and
customization of functional and conventional yarns, it is possible to strategically
manipulate the visual appeal and structural characteristics, as well as optimize the
electrical and mechanical attributes of a sensing textile. In this study, we present
a novel approach for the customization and formation of three-dimensional piezo-
resistive textile using thermoforming principles and the utilization of melting yarns.
The outcome is a durable textile structure and close integration, which effectively
reduces sensor drift and enhances accuracy, all while prioritizing comfort.

The utilization of digital knitting techniques allows for the production of pressure-
sensitive textile interiors and wearables that transition from two-dimensional to three-
dimensional forms. Notable examples include an intelligent mat measuring 45 ×
45 cm, which incorporates 256 pressure-sensing pixels, as well as a form-fitted
shoe that is circularly knitted and features 96 sensing pixels distributed across its
three-dimensional surface. In addition, a visualization tool and framework have been
developed to process spatial sensor data as image frames. The convolutional neural
network (CNN) models developed in our study demonstrate the ability to accurately
classify seven fundamental activities and exercises, as well as seven distinct yoga
poses, in real time. The achieved accuracy rates for these classifications are 99.6%
and 98.7%, respectively. Additionally, we showcase the utilization of our technology

© The Author(s), under exclusive license to Springer Nature Singapore Pte Ltd. 2024   43
M. H. Akhtar and J. Ramkumar, *AI for Designers*,
https://doi.org/10.1007/978-981-99-6897-8_3

across diverse domains, encompassing rehabilitation and sport science, as well as wearables and gaming interfaces [2].

Case study 2: Wearable reasoner: Towards enhanced human rationality through a wearable AI assistant [3]

In this study, media lab at MIT [2] introduces the "Wearable Reasoner," a prototype wearable device designed to assess the presence or absence of supporting evidence in arguments. The primary objective of this system is to encourage individuals to critically evaluate the justifications behind their own beliefs as well as the arguments put forth by others. In this experimental study, we investigated the effects of argumentation mining and the explainability of AI feedback on users. This was done through a verbal statement evaluation task. The findings of this study indicate that the utilization of a device equipped with explainable feedback yields positive outcomes in promoting rationality among users. This is achieved by facilitating the distinction between statements that are substantiated by evidence and those that lack evidential support. Users demonstrate a greater inclination towards accepting claims that are accompanied by reasons or evidence when they receive assistance from an AI system that provides explainable feedback. The findings from qualitative interviews indicate that users engage in internal processes of reflection and integration when considering new information for their judgment and decision-making. Participants expressed satisfaction with the presence of a second opinion, highlighting the enhanced evaluation of the arguments presented.

The results of our study indicate that the inclusion of explainable AI feedback in our prototype has a notable impact on users' agreement levels and their perception of the presented arguments' reasonableness. When provided with such feedback, users tend to exhibit a higher level of agreement with claims that are accompanied by supporting evidence and perceive them as more rational in comparison to claims that lack such evidence. During qualitative interviews, users provide insights into their cognitive processes of speculation and integration of information derived from the AI system into their own judgment and decision-making. This leads to enhanced evaluation of the arguments presented.

Case study 3: Children creativity with robots [2]

The creative capacity of children, characterized by their ability to generate original, unexpected, and valuable concepts, has been recognized as a significant factor in their educational achievements and individual development. Research on standardized methods for assessing creativity and divergent thinking has indicated that there is a decline in these cognitive abilities as children transition into elementary school, with a notable increase in convergent thinking, particularly during the fourth grade. One contributing factor to this phenomenon is the increasing rigidity of school curricula, which often neglects the element of creative play. The rise of artificial intelligence poses a particular concern for children who are growing up during this era. This is due to the fact that jobs that involve mechanical and repetitive tasks, which traditionally require structured thinking, are increasingly being automated and delegated to machines. In order to achieve success in the realm of intelligent agents, it is imperative that we provide children with the necessary tools to comprehend the functioning of these agents. Additionally, we must foster their ability to engage in creative thinking, enabling them to collaboratively generate innovative artifacts alongside these agents. This process necessitates the cultivation of imaginative and original thought. This study investigates the potential effectiveness of utilizing social robots as a means to enhance children's creative thinking abilities. We propose two methods by which robots employed as pedagogical instruments can enhance children's creative thinking abilities: firstly, by demonstrating artificial creativity, which involves showcasing the application of innovative concepts; and secondly, by providing creativity scaffolding, which entails posing reflective inquiries, acknowledging novel ideas, and fostering creative conflict [2].

We have developed a set of four interactive game-based activities that facilitate collaborative engagement between children and robots, while also providing opportunities for diverse modes of creative expression. The Droodle Game is a recreational activity that promotes verbal creativity. Magic Draw, a tool that facilitates the expression of figurative creativity. The WeDo Construction program, in conjunction with Jibo, provides opportunities for fostering construction creativity. Additionally, the Escape Adventure program facilitates the development of divergent thinking skills and enhances creative problem-solving abilities. The behavior of the robot was intentionally designed to serve two purposes: either to provide scaffolding for the child's creative thinking process or to simulate human creativity through artificial means. We conducted a study to assess the impact of social robots on children's creativity. This was achieved by conducting comparative analyses of children engaging in creativity games while interacting with a robot that exhibited creativity-inducing behaviors (referred to as the creative condition) and children engaging in the same games without exposure to creativity-inducing behaviors (referred to as the non-creative condition). The findings of the study indicate that children who engaged with the creative robot demonstrated elevated levels of creativity compared to children who interacted with a control robot that lacked creative capabilities. It is concluded that children have the ability to imitate and replicate the creative expression of social robotic peers through social emulation. When provided with scaffolding to enhance creativity, children demonstrated elevated levels of creative thinking. The implementation of this approach has facilitated the development of a scaffolding paradigm for robots, which effectively promotes creativity among young children. This project offers design guidelines for interactions between children and robots, with a focus on fostering creative thinking. It also presents empirical evidence supporting the notion that social robots, through their display of creativity-inducing behaviors, can effectively enhance creativity in young children [2].

What is artificial intelligence?

It is the science and engineering of making intelligent machines, especially intelligent computer programs. It is related to the similar task of using computers to understand human intelligence, but AI does not have to confine itself to methods that are biologically observable. Yes, but what is intelligence? Intelligence is the computational part of the ability to achieve goals in the world. Varying kinds and degrees of intelligence occur in people, many animals, and some machines. Isn't there a solid definition of intelligence that does not depend on relating it to human intelligence? Not yet. The problem is that we cannot yet characterize in general what kinds of computational procedures we want to call intelligent. We understand some of the mechanisms of intelligence and not others.

Artificial intelligence (AI) has become an overhyped buzzword in numerous areas, including the design industry. Conversations are happening between designers and developers over the future influence of AI, machine learning, and deep learning. So, what does design contribute to the discussion? With AI, new customer–product partnerships will need to be developed. These interactions will be the start of an ongoing dialogue between businesses and consumers over what artificial intelligence can and should accomplish for products and services. As a result of designers providing the required empathic framework for innovation, an AI-powered firm will be successful.

When it comes to new product development, AI and machine learning (ML) are helping to expedite the process for everyone from startups to large corporations. On Indeed, LinkedIn, and Monster, there are 15,400 open openings for DevOps and product development engineers with AI and machine intelligence. From $519 billion to $685 billion is expected to be the size of this year's connected product market harnessing the potential of AI and ML, according to Capgemini. AI-based apps, goods, and services are rapidly advancing, and this will cause the IoT platform market to consolidate as a result. The IoT platform providers who focus on solving business problems in specific vertical markets have the best chance of surviving the impending shakeout of the IoT platform market. The IoT platforms and ecosystems enabling smarter, more connected products must prepare now for how they will stay up with AI and ML becoming increasingly integrated in new product development. It will be impossible to keep up with the rapidity of change if we solely rely on technology, as many IoT systems do now [4].

How AI and machine learning accelerate product development workflows in manufacturing?

According to a survey conducted by MIT Technology Review Insights in 2020, manufacturing is among the top two industries embracing AI. AI and ML deliver numerous benefits to manufacturing use cases, such as product research, development, and production, inventory management, process and quality control, and predictive maintenance. Advanced AI technologies are currently being included into the workflows of industry leaders. Using NVIDIA software libraries and the NVIDIA EGX platform for accelerated computing, Foxconn Group, for example, has integrated AI for automated high-precision inspection of its products' components and tools [5].

AI in product development is much more than a mere add-on to accelerate procedures. It is important to keep in mind that merely adopting AI does not guarantee the desired outcomes. You must adopt a data-strategy-first mentality to realize its full potential. The objective here is to properly consider, generate, and plan how the AI solution will support your organization. By taking a well-considered approach to AI, you can do more than just ensure that your business stays current and in step with market trends. It is about utilizing these technological breakthroughs to manufacture items faster than your competitors to maintain a competitive advantage. Although it may make sense to imitate what other prominent companies are doing, there is no single technique that is optimal for all businesses. Therefore, be mindful of common AI adoption errors. The key to installing the appropriate system(s) for your organization is a clear approach [5].

The development of AI and ML algorithms has made it easier for businesses to support their product design and improvement centers. Businesses are using a variety of AI and ML techniques to create and market new products as well as to enhance the ones they already have. Unlike conventional data collection methods, AI and ML may offer organizations the finest potential solutions. At the same time, AI promotes quicker product creation and helps achieve competitive advantages. Thus, it is the need of the hour that different stakeholders use and leverage the potential of AI in product design and development [6].

Generative design

The concept of generative design refers to a computational approach that utilizes algorithms and data-driven processes to generate and optimize design solutions. Generative design entails the utilization of a computer program to produce diverse outcomes that align with the predetermined criteria of the respective organizations. In order to produce a generative design, designers and engineers provide design objectives and parameters as input. The aforementioned factors encompass various aspects such as material selection, manufacturing techniques, financial limitations, and alternative design possibilities. Subsequently, the software utilizes machine learning algorithms to acquire knowledge from each iteration and comprehend the underlying concept. The artificial intelligence (AI) and machine learning (ML) algorithms subsequently produce a diverse range of alternatives. The technology offers a means of enhancing human labor by creating a virtual replica [6].

Build virtual copy

Artificial intelligence (AI) plays a significant role in the development of a digital replica of a tangible manufacturing system. This encompasses both an individual machinery asset as well as an entire machinery system. Digital copies play a crucial role in facilitating real-time diagnosis, process evaluation, prediction, and visualization of product performance. Data science engineers employ artificial intelligence (AI) and supervised machine learning (ML) algorithms to aid in the manipulation and analysis of these digital replicas. The entire process involves the analysis of historical records and unannotated data collected through real-time monitoring. Virtual models play a crucial role in enhancing production scheduling, quality enhancement, and maintenance processes [6].

Assembly line optimization

Optimization of Assembly Line Efficiency artificial intelligence (AI) plays a crucial role in the enhancement of product development through the optimization of assembly lines. Let us examine an illustrative scenario wherein the occurrence of fatigue symptoms in an equipment operator prompts the receipt of notifications by both supervisors and technicians. Hence, the system initiates contingency plans and other reorganization activities automatically [6].

Predictive maintenance

The primary benefit of artificial intelligence lies in its capacity for forecasting and prediction. AI technologies play a crucial role in the identification of potential instances of downtime and accidents through the analysis of real-time data. Artificial intelligence (AI) facilitates the ability of technicians and supervisors to predict an impending failure in advance of its actual occurrence. Therefore, artificial intelligence (AI) has the potential to enhance development efficiency through the reduction of machine failure costs [6].

Inventory management

Machine learning (ML) and artificial intelligence (AI) algorithms facilitate the optimization of planning activities. Artificial intelligence (AI) tools have been found to yield results that are more precise and reliable compared to conventional methods. Effective inventory control plays a crucial role in mitigating potential issues such as cash-in-stock and out-of-stock situations. Inventory management is commonly known as stock management in academic literature. Inventory management encompasses the systematic activities involved in the organization, procurement, commercial transactions, and upkeep of a company's inventory [6].

Quality assurance

Quality assurance (QA) is a systematic process that aims to ensure the quality and reliability of products or services. It involves the implementation. Quality assurance is a final stage implemented to ensure the quality of products and services. The purpose of this test is to assess the quality of the product provided by the company, thereby evaluating its adherence to established standards. Historically, the responsibility of quality assurance has been entrusted to proficient individuals and has predominantly been a labor-intensive task. Currently, artificial intelligence (AI) and image processing algorithms possess the capability to technically ascertain whether an item has been developed and produced in accordance with the correct standards. The implementation of a real-time and automated verification system has facilitated the product development process [6].

Quick decision-making

Decision-making can be defined as the cognitive process of selecting the most appropriate combination from a range of available options. Artificial intelligence (AI) and machine learning (ML) algorithms are instrumental in facilitating prompt decision-making regarding various products, surpassing the effectiveness of conventional methods [6].

Process optimization

Process optimization is primarily undertaken to establish the long-term viability of a product within the marketplace. Artificial intelligence (AI) software plays a crucial role in assisting organizations in optimizing their production and development processes across all levels, with the ultimate goal of achieving sustainable production. The process aids in the identification of operational bottlenecks and subsequently eliminating them [6].

AI in new product development

Startups' ambitious AI-based new product development is driving AI-related investment with $16.5B raised in 2019, driven by 695 deals according to PwC/CB Insights MoneyTree Report, Q1 2020. AI expertise is skill product development teams which are ramping up their recruitment efforts to find, with over 7800 open positions on Monster, over 3400 on LinkedIn, and over 4200 on Indeed as of today. One in ten enterprises now uses ten or more AI applications, expanding the Total Available Market for new apps and related products, including chatbots, process optimization, and fraud analysis, according to MMC Ventures [4].

A notable finding reveals that 14% of enterprises, which are at the forefront of utilizing artificial intelligence (AI) and machine learning (ML) for the purpose of new product development, generate over 30% of their total revenues from fully digital products or services. Furthermore, these enterprises exhibit superior performance compared to their counterparts by effectively employing nine essential technologies and tools. According to a study conducted by PricewaterhouseCoopers (PwC), it was discovered that organizations identified as Digital Champions exhibit a notable advantage in terms of revenue generation from novel products and services. Furthermore, it was observed that a considerable proportion of these champions, specifically 29%, manage to generate over 30% of their total revenues from new products within a two-year timeframe. Digital champions hold high expectations regarding the potential for attaining increased advantages through personalization as well [4].

A significant proportion of advanced enterprises, specifically 61%, that employ artificial intelligence (AI) and machine learning (ML) techniques, commonly referred to as Digital Champions, have adopted fully integrated Product Lifecycle Management (PLM) systems. In contrast, a mere 12% of organizations that do not currently utilize AI/ML technologies, known as Digital Novices, have implemented such integrated PLM systems. Product development teams that have made significant advancements in their utilization of artificial intelligence (AI) and machine learning (ML) are able to achieve notable improvements in economies of scale, efficiency, and speed gains across the three fundamental areas of development. Digital Champions primarily focus on acquiring time-to-market and speed benefits in various domains such as Digital Prototyping, Product Lifecycle Management (PLM), collaborative development of novel products with customers, Product Portfolio Management, and the implementation of Data Analytics and artificial intelligence (AI) [4].

Artificial intelligence (AI) is currently being utilized in the strategic development, execution, and optimization of interlocking railway equipment product lines and systems. The implementation of engineer-to-order product strategies results in a significant proliferation of product, service, and network alternatives, exhibiting an exponential growth pattern. The process of optimizing product configurations necessitates the utilization of an artificial intelligence-based logic solver capable of considering all relevant constraints and generating a Knowledge Graph to facilitate the deployment process. The utilization of AI by Siemens to identify the most optimal configuration from a pool of 1090 potential combinations offers valuable insights into the potential of AI in facilitating large-scale new product development [4].

The initial step towards overcoming obstacles in the successful launch of new products involves the utilization of artificial intelligence (AI) to enhance the precision of demand forecasting. Honeywell is employing artificial intelligence (AI) technology in order to mitigate energy expenses and mitigate adverse price discrepancies. This is achieved through the monitoring and analysis of price elasticity and price sensitivity. Honeywell is effectively incorporating artificial intelligence (AI) and machine learning algorithms into their procurement, strategic sourcing, and cost management practices, resulting in significant benefits throughout the new product development process [4].

The utilization of artificial intelligence (AI) methods is employed to develop and optimize propensity models. These models are utilized to determine the most financially advantageous cross-selling and up-selling opportunities for product line extensions and additional products. This is done by considering various factors such as product line, customer segment, and persona. Propensity models are frequently employed by data-driven teams in the realms of new product development and product management to delineate the products and services that possess the greatest likelihood of being acquired. Frequently, propensity models rely on imported data and are constructed within the Microsoft Excel platform, resulting in a laborious ongoing utilization process. Artificial intelligence (AI) is enhancing the efficiency of the development, optimization, and financial impact of up-selling and cross-selling strategies through the complete automation of these processes [4].

Artificial intelligence (AI) is facilitating the development of advanced frameworks that effectively decrease the time required to bring products to market. Simultaneously, these frameworks enhance the quality of products and provide increased flexibility to accommodate individualized customization requests from customers. Artificial intelligence (AI) is facilitating improved coordination among various stakeholders, including suppliers, engineering teams, DevOps, product management, marketing, pricing, sales, and service. This enhanced synchronization increases the likelihood of a new product achieving success in the market. Prominent entities in this domain encompass the Autonomous Digital Enterprise (ADE) developed by BMC. The ADE framework developed by BMC demonstrates the potential for growth-oriented organizations to implement advanced business models that incorporate AI/ML capabilities. This framework allows businesses to effectively manage and transform their operations, while also providing competitive advantages through increased agility, customer focus, and the ability to generate actionable insights [4].

The utilization of artificial intelligence (AI) for the purpose of analyzing and delivering recommendations pertaining to the ongoing enhancement of product usability is on the rise. It is a prevalent practice within the fields of DevOps, engineering, and product management to conduct A/B tests and multivariate tests in order to ascertain the user preferences for usability features, workflows, and application and service responses. Drawing from personal experience, a notable difficulty encountered in the process of developing new products lies in the task of crafting a user experience that is both efficient and captivating, thereby transforming usability into a prominent advantage for the product. The integration of AI techniques into the fundamental process of developing new products enables the attainment of enhanced usability and the provision of pleasurable customer experiences. Rather than perceiving a novel application, service, or device as burdensome to operate, artificial intelligence (AI) has the potential to offer valuable insights that can enhance user experience by making it more intuitive and enjoyable [4].

The application of artificial intelligence (AI) in the domain of demand forecasting for novel products, encompassing the identification of primary causal factors that significantly influence new sales, is currently yielding promising outcomes. The process of forecasting demand for a next-generation product exhibits significant variation among companies, stemming from a range of pragmatic methodologies. These methodologies encompass soliciting input from channel partners, as well as both indirect and direct sales teams, and employing advanced statistical models to estimate the quantity of units that will be sold. Artificial intelligence (AI) and machine learning (ML) have demonstrated their significance in analyzing previously unknown causal factors that impact demand. The utilization of artificial intelligence (AI) in the design process of forthcoming Nissan automobiles has resulted in the optimization of new product development, leading to a reduction in the duration of new vehicle development schedules by several weeks. The pilot program implemented by Nissan to expedite the development of new vehicle designs through the utilization of artificial intelligence is referred to as DriveSpark. The program was initiated in 2016 as an experimental endeavor and has subsequently demonstrated its efficacy in expediting the development of new vehicles, while also ensuring adherence to compliance and regulatory standards. Furthermore, artificial intelligence has been employed to prolong the lifespans of preexisting models [4].

3.2 Human-Centered AI

Machine learning is an extremely useful technique that powers many different features, such as personalized content recommendations and enhanced user interfaces. Machine learning is becoming increasingly integrated into everyday product interactions, with Apple, Google, Facebook, and Amazon as the leaders. Because of this, designers of digital products should educate themselves in the field of machine learning. Although mastering machine learning is likely to be a specialized skill set for designers soon, they will soon be unable to function without it. There are numerous publications that adequately define machine learning. Designers who put off learning ML are being warned that they will fall behind the curve. The relationship between design and machine learning is like a flywheel; each discipline strengthens the other. In tandem, they make possible cutting-edge consumer and commercial possibilities.

What can design contribute to machine learning?

The quality of data used by machine learning is improved with the aid of design. Learning by machine consumes a lot of resources. Learning algorithms perform best when fed massive volumes of comprehensive data free of biases and other potential sources of error. For Spotify's song recommendations to be effective, for instance, the service's algorithms must have information about the kind of music that customers prefer [7]. If the app's browsing experience is clumsy, users will be more likely to listen to the first few songs they see, regardless of whether they like them. Recommendations are less reliable when "noisy" indications are present. For ML-powered applications, designers can help to improve accuracy and efficiency by crafting experiences that filter out irrelevant information.

How design aids in establishing user expectations and building trust?

Twitter [8] relies heavily on a variety of ML algorithms. By using an algorithm, the "top tweets" feed determines the order in which tweets are displayed. The algorithms determine how often you see tweets that your friends like, retweet, or reply to. Trending topics displayed in the sidebar are selected by a separate set of algorithms. It's not uncommon for Twitter users to wonder what's going on. Marketers in the social media industry have produced lengthy articles detailing how to game Twitter's system. Frustration and distrust stem from misunderstandings and lack of clarity. Clarity can be added to ML-powered interfaces by designers, notably user experience writers, the unsung heroes of usability. The developers of Twitter have added some context if a tweet from someone you don't follow appears in your timeline: liked by Matthew, replied to by Raquel, retweeted by Freyja, etc. With the headline "Trends for you," the trends in the sidebar make it clear that they have been selected just for you.

User confidence and familiarity are bolstered by these thoughtful touches. The success of ML depends heavily on the confidence of its users. The People + AI Research [9] (PAIR) initiative at Google has published extensive guidelines for establishing credibility in ML-powered apps. Understanding the correct amount of detail to explain AI-driven systems is crucial for users, as these systems are built on chance and uncertainty. When users have accurate mental models of the system's strengths and weaknesses, they know when and how to rely on it to help them reach their objectives. In a nutshell, the ability to provide an explanation is directly related to gaining people's trust.

What benefits does machine learning hold for creative work?

Questions concerning user behavior can be answered by machine learning. To construct and iterate on user interfaces, designers frequently need to make educated guesses about their target audience. In most cases, the most difficult assumptions take the form of "if a user does this, then that will happen." In other words, if a customer uses a discount code during checkout, they are more likely to buy something. Despite the significant impact they have on the value of a company, these assumptions are generally the result of convoluted statistical research and, unfortunately, educated guesswork. This is where machine learning comes in, since it can assist simplify the process by creating complicated models of user behavior that are straightforward to test. A Bayesian network [10] is a type of this method. What will happen if a user does (or doesn't) do something is an example question that can be answered by a Bayesian network. Engineers and architects can use machine learning in this way to make better informed judgments with less guesswork. Thanks to machine learning, user interfaces may be tailored to everyone.

How amazing would it be if a company could make the experience of using their app totally different for each individual user? If this company must design this for a specific user at a specific time, how would they adjust the experience to meet the user needs? It's unsettling to imagine that designers would have to individually adjust the placement of buttons and menus for each user to make an app more intuitive. Yet, in 2018, Facebook [11] started arranging the app's navigation bar according to the functions its users access more often, an example of ML's practical application.

A look into the future of machine learning and design

The design and machine learning hamster wheel has only just begun to turn. The rapid pace of progress in machine learning makes it challenging for many designers to catch up with the subject. The pace of invention will quicken once machine learning is incorporated into standard design education. The ability of Sketch, Figma, or Adobe XD to foretell a mockup's usability based on actual user data would be incredible. With the data collected by tools like FullStory [12], you may teach an ML algorithm to behave like your consumers, even when presented with novel displays or processes. If you'd prefer, you may investigate a solution that will automatically design the UI for your app depending on some general settings. To achieve the desired result, a designer may only need to alter a few sliders (such as session time, CTA conversion rate) and an algorithm will make the necessary adjustments. A technology of this type might adapt the user interface (UI) in real time, during production, based on how users are really interacting with the system. Blue-sky thinking like this is essential in the development of new products. Designers are in a prime position to test the limits of ML and demonstrate how algorithms can provide tangible value to end users and organizations.

A peek into the philosophy of Google

To what extent do designers leverage the potential of artificial intelligence (AI) and enhance its accessibility?

In contemporary times, it is imperative for product developers to contemplate a factor that was not previously relevant to their predecessors or even their own past selves: the potential of artificial intelligence to offer an innovative resolution to the given problem. The response is increasingly leaning towards affirmation, although it is important to acknowledge that AI is not a cure-all solution. Instead, when employed judiciously, AI has the potential to enhance experiences by offering individuals unique predictive insights, personalized services, and a deeper understanding of their own needs. This technology holds potential for individuals in creative fields; however, its emergence also raises numerous inquiries.

The question at hand pertains to the categorization of artificial intelligence (AI) as either a resource, an instrument, or a combination of both. How can individuals enhance their AI literacy in order to ensure that algorithmic decision-making produces outcomes that are relevant to all users?

In relation to this matter, recent resources such as the People + AI Guidebook by PAIR and the Material Design principles for the ML kit API by Google may be beneficial. Rachel Been, the creative director of Material Design, elucidates that the framework is being established to facilitate the comprehension of this novel technology by our users. Nevertheless, the development of such a framework necessitates a meticulous and nuanced approach that is firmly rooted in the needs and demands of individuals. In order to gain a deeper understanding of the methods by which designers can effectively harness and imbue artificial intelligence (AI) with human-like qualities, we conducted an analysis of a brief conversation involving Rachel, a design manager at DeepMind Health, and Jess, a PAIR lead and cofounder of the People + AI Guidebook.

What is your interpretation of the term "human-centered AI?"

The individual under discussion is Jesse Holbrook. Artificial intelligence (AI) is currently regarded as a state-of-the-art technological tool. The current stage of development in relation to personal computers, the World Wide Web, and mobile devices is still in its early phases. However, similar patterns that were observed during the emergence of these technologies can be identified at present. Individuals often engage in the exploration of novel endeavors, wherein their resultant product exhibits commendable qualities. However, upon reflection, it becomes apparent that the experiential creation fails to satisfy an authentic human desire. The objective of this study is to emphasize the importance of focusing on individuals as the primary catalyst for achieving success. Prioritizing the well-being and interests of individuals is consistently advisable when engaging in any form of exploration, product development, or scholarly investigation.

Rachel Been is an individual whose identity is known. When developing artificial intelligence systems with human users as the target audience, it is crucial to take into account the challenges associated with managing the inherent uncertainty that arises from machine learning and artificial intelligence. Designers must possess adaptability in order to effectively address inquiries such as, "What would happen if the user encounters an error?" What if she requests additional information regarding the actions of the artificial intelligence? How can one facilitate the process of familiarizing her with the subject matter in order to enable her to develop a comprehensive understanding and appreciation of it? Traditional design patterns commonly exhibit a sequential progression from one phase of the user experience to the subsequent phase. The implementation of artificial intelligence necessitates the establishment of a novel set of criteria.

Jess: Indeed, it is possible to disregard the human-scale cause-and-effect correlations when engaging with artificial intelligence. Nevertheless, a fundamental concern in the design of AI systems centered around human interaction is the challenge of elucidating the functioning of this technology, which operates on a scale surpassing human capabilities, in a manner that is comprehensible to humans.

Oznur: One intriguing inquiry pertains to the manner in which we opt to elucidate the technology. In the initial stages of our engagement with artificial intelligence (AI) and machine learning (ML), there was a prevailing assumption that providing individuals with transparent information regarding the origins of data and the resultant computational processes would be imperative in persuading them to adopt this technology. It has been observed, however, that the level of mathematical proficiency possessed by a user is not a prerequisite for accepting the conclusions of an algorithm, provided that the algorithm is capable of articulating its logical reasoning to the user. For example, the machine learning models that have been developed for digital imaging have the capability to identify lesions and other abnormalities in the eye. Subsequently, these models can offer clinical advice based on their diagnostic findings. The outcome remains comprehensible irrespective of one's knowledge regarding the calculation methodology employed. With the aforementioned objective in mind, our intention is to compose a piece of writing that exhibits characteristics more akin to a narrative rather than a user guide, thereby facilitating the user in making informed decisions.

Rachel: During the process of designing patterns for ML Kit's demo experience, we encountered instances where the API's object detection functionality, which employs visual search to ascertain the nature of an object, would promptly and accurately identify the object. Due to the product's swift functionality, the user encountered significant difficulties. This raises the quandary of whether it is preferable to decelerate the pace of the activity in order to allow the user sufficient time for cognitive processing, or whether it is more advisable to align with their preexisting conceptions regarding the functioning of computation. One effective approach to addressing this situation involves refraining from excessive veneration of technology and instead acknowledging its instrumental role in achieving a more gratifying user experience.

The concept of perceiving AI as a tool encapsulates a significant transformation for designers, namely the integration of AI as a fundamental component of the design process. What is your interpretation of the given information?

Rachel: I have been contemplating whether artificial intelligence (AI) should be classified as a material entity or a utilitarian instrument. Due to its capacity to retain and recall user inputs, the system exhibits adaptability and can be molded to suit the user's preferences akin to a malleable substance. Moreover, the application of AI is also influencing the development of front-end user experiences.

Oznur: An example of AI implementation can be observed in the construction sector, where our research team focuses on the development of algorithms specifically designed to enhance the ability to predict health-related concerns. Nevertheless, when a corporation such as Google Photos integrates artificial intelligence (AI) into the search functionality of their product to create an innovative categorization mechanism for users' photographs, the AI assumes a role akin to a tool employed by individuals.

Jess: In the realm of materials, it is frequently observed that discernible limitations exist, which become evident upon the occurrence of failure. Label makers allow users to create white letters by punching them out of plastic sheets. The employed failure case involves exerting excessive pressure on another material in order to construct the interface, thus rendering it a noteworthy illustration of leveraging the limitations of a medium in the context of design. Within the realm of artificial intelligence, there exists an ongoing discourse regarding the most effective approach to elucidate the constraints of a particular material in order to enhance public understanding of its merits.

Rachel: Experiments involving the generation of music or an entire science fiction movie through machine learning techniques have already facilitated the exploration of limitations within machine learning for compelling purposes, particularly within the domain of artistic expression. The Uncanny Valley effect is encountered due to limitations imposed by technology, yet it is this very phenomenon that captivates us as observers. The imposition of such limitations can be particularly problematic in utilitarian contexts, as exemplified by the potential challenges faced by individuals like Oznur in the healthcare sector, or when our efforts to provide assistance in sensitive matters are hindered. In the context of health care, the imposed limitations may not be optimal; however, in the context of a game or an art project, they can be highly enjoyable.

In the realm of design, striking a balance between experimentation and the establishment of standards that facilitate a common language among designers is a pertinent concern. How can one effectively navigate this tension?

Rachel: The implementation of standardized practices within the system will facilitate the consolidation of our resources and enable their further development. The People + AI Guidebook can be characterized as a compendium of recommendations rather than a rigid set of principles. It occupies a middle ground between prescriptive tactical guidance and abstract philosophical suggestions. Nevertheless, it establishes the foundation for novel concepts. Despite significant advancements in hardware, data, or augmented reality, there remains a fundamental mental model that allows for comprehension and interpretation of any radical changes that may occur in the front end of an application or service.

Jess: Designers ought to possess the capability to peruse this manual during the morning hours, incorporate its recommendations into their workflow later in the day, and experience a sense of accomplishment. The current discourse surrounding AI revolves around its financial and ethical ramifications. However, irrespective of one's underlying motivations, it is imperative to prioritize the human-centric design of AI systems. Hence, the concept of direction functions as the cohesive element that unifies the aforementioned discussions and available resources.

Oznur: Indeed, rather than grappling with the complexities of an indeterminate terrain, one can direct their attention towards operating within a clearly delineated structure. The development of principles, such as explainability, holds significant importance in the context of leveraging user feedback. Designers often encounter limitations in their profession.

Jess: Limitations are consistently embraced.

Oznur: However, it should be noted that this item does not consist of a collection of adhesive labels. To effectively apply the advice provided, it is necessary to engage in cognitive processing in order to ascertain its applicability within one's professional routine. These insights have emerged from a collective amalgamation of individuals operating within this particular industry. The most pleasurable aspect lies in extracting relevant insights from this information and adapting them to suit one's individual circumstances.

To what extent can the future-proofing of guidance be achieved given the dynamic nature of AI? Is it possible to accomplish that task?

Rachel: The process of establishing standards for the standardization of a digital button differs significantly from the formulation of criteria for artificial intelligence (AI), which undergoes continuous evolution. Therefore, it is incumbent upon designers to proactively caution users, making a clear distinction between practical recommendations that are firmly established through extensive research and standardization, and guidelines for AI systems that are subjected to change due to the rapidly evolving nature of this field.

Jess: It has consistently been emphasized that our intention is not to transform Google into the sole global authority. The proposed strategy involves the widespread distribution of this information to a subsequent cohort of one thousand specialists, who will subsequently impart this knowledge to an additional ten thousand individuals, and so forth. It is imperative to acknowledge that this constitutes a pivotal component of said procedure.

Some fundamental terms to learn for designers [13]

We have an obligation as UX team members to learn how our apps and websites are constructed from the ground up. While artificial intelligence (AI) is not a new field in computer science (was formalized in the 1950s), it is a new frame of reference for those of us performing UX design and content marketing for consumer products. Google is dedicated to explaining the inner workings of artificial intelligence systems in simple terms for the benefit of both its users and the wider public. Expectations of how the same AI terminology might be used or applied will vary greatly depending on whether you are a beginner, expert, researcher, developer, designer, content strategist, policy advocate, or any of the other ten categories listed above. When we all have the same definition for a term in the same context (in the same dictionary), we can communicate effectively. Clearly and simply explaining how AI systems function can assist UX teams engage with users, allowing them to gain trust in those systems while also addressing users' demands.

Primordial vocabulary

We analyzed the first batch of articles in our People + AI Research collection to determine the most frequently used AI terms and use them as the basis for a core vocabulary for UX professionals. For more context, we polled a large sample of Google's designers on their thoughts about ML and how they may define it. From this, we were able to create a foundational AI vocabulary list, from which we extracted six terms utilized (and frequently misunderstood) by UX designers, researchers, and content strategists. In this article, we provide concise explanations of each.

Artificial intelligence

AI refers to the study of making machines intelligent so that they may learn to spot patterns and become effective at assisting humans in tackling difficult problems. When a computer decides based on a forecast, whether by using simple rule-based systems or heuristic methods like "if rain, then umbrella," the computer is using artificial intelligence. However, with machine learning (discussed further down), the rules for making choices are taught.

Machine learning (ML) is a branch of AI that deals with developing methods and strategies to train computers to perform tasks without being explicitly instructed to do so. It's possible to teach a computer to learn in a variety of ways. The most important of these is supervised learning, in which a computer is taught to anticipate outcomes from a large dataset, such as your commute time. We'll save discussion of the other well-known methods, such as unsupervised, semi-supervised, and reinforcement learning, for another time (or you can learn the technical details on your own with our Machine Learning Glossary for developers).

Machine learning model

Combinations of various specialized mathematical operations that are all interrelated when put together, they stand in for the process an artificially intelligent system will go through to make a call. An ML model can predict when you might arrive at your destination based on your historical travel habits and current conditions. The terms "algorithm" and "neural network" are sometimes used interchangeably with "ML model." Neural networks are simply one type of ML model; algorithms, on the other hand, are more generic, almost recipe-like computing techniques. Neural networks are so named because they resemble brain cells, or neurons. In humans, the ability to learn, create, and conceptualize is all down to the nerve cells called neurons, which carry electrical impulses.

Classification

Predicting to which of several predefined classes a given input likely belongs.

An ML model is continually working in the background to determine if an incoming email is spam or not (and if there is any doubt, Gmail will ask you to verify the sender's email address). Although this type of binary prediction works well for answering yes/no questions, classification models are capable of much more. For a given input, these models can make predictions over several different classes. A model like this might determine that an email is "not spam" and "important," with the tags "financial" and "follow-up," respectively. Try out our What-If Tool for a taste of the flexibility and power of classification. Teams can assess the outcomes of

ML models with no coding required thanks to the What-If Tool, an interactive data visualization.

Regression

To ask a model to make a numerical prediction for a certain circumstance. A "regression" operation is performed by a model when you query for a flight's price two weeks from now. The ML model needs to provide more nuanced feedback than just a simple yes/no for this user experience to succeed. To provide more nuanced information, the prediction is based on historical data presented as continuous numerical values.

Let's say you're tasked with creating a window display for winter apparel and accessories. Classification might be compared to organizing winter hats and scarves into separate containers. To do this, we'll think about the various ways in which you and your consumers identify scarves, hats, and other headwear based on their shape. Consider a complex regression problem as the equivalent of creating a window display for a winter clothing store that features stylish yet functional winter coat options appropriate for the area's typical weather conditions. Adjust your selection of scarves, hats, earmuffs, socks, fleece layers, and coats according to the season, the weather, your clients' needs and wants from previous years, and the current fashion trends. Incredibly sophisticated user experiences are powered by regression predictions, which can be used to do everything from predict fluctuations in currency prices to score songs for a curated playlist to evaluate image quality. Determine whether a regression model is useful for your users by thinking about how much depth and complexity they need in the final product or service. Explore how the Google Clips team implemented a regression model to create a fully automated camera.

Percentage of certainty; a quantifiable representation of reassurance

Humans use phrases like "I think this person may be 35 years old" when attempting to estimate a person's age. Words like "I suppose" and "may be" indicate a lack of confidence or conviction, so we know it's just a hunch. In a similar vein, you can consider model forecasts to be educated approximations seasoned with some uncertainty. For example: "I am 73.3% confident that this person is 35 years old," where 73.3% represents the model's level of assurance (or doubt, if you're a glass-half-empty type of person). Product teams will use the confidence level to determine if an answer is satisfactory. The models we employ tend to be rather accurate, so if they predict a 70% chance of rain for today, we may feel comfortable suggesting that our users bring umbrellas to work.

3.3   Artificial Intelligence Principles

Are principles of artificial intelligence (AI) significant? What are the fundamental principles of artificial intelligence? [14]

The objective is to develop an AI product of exceptional quality, and the initial step involves discussing it with colleagues. The endeavor to garner unanimous support encounters the challenge of divergent interpretations of the term "best" among individuals, thereby raising concerns about potential customer responses to certain ideas. This issue persists despite one's well-meaning intentions.

When customers inquire, what do you direct their attention towards? How can one ensure that their products do not have detrimental effects on the overall sustainability and viability of their company?

To achieve progress, it is imperative to employ a strategic instrument that propels the organization's operations. These entities are commonly referred to as principles, guidelines, company charter, and values within the academic discourse. There exists a necessity to establish a methodology for generating ethical guidelines, as well as implementing mechanisms to ensure compliance within the organization. From a user experience (UX) perspective, the formulation of these AI principles serves to guide the establishment of objectives and the selection of metrics for their evaluation. The majority of the AI principles discussed in this episode pertain to lofty and ambiguous concepts. The field of machine learning is currently experiencing rapid advancements and developments. Developing rigid guidelines that remain relevant beyond a six-month timeframe would be an exceedingly challenging task. Conversely, what strategies can be employed to avoid excessive ambiguity or reliance on superficial marketing language that lacks substantive meaning? The recommended course of action is to execute or apply the proposed measures. If the AI principles can be effectively implemented, then they possess a sufficient level of definition to be adhered to.

Hagendorff [15] conducted a comparative analysis of the principles adopted by various associations and companies. Several common principles are commonly employed by most companies, including privacy protection, accountability, fairness, non-discrimination, justice, transparency, openness, safety, cybersecurity, common good, sustainability, well-being, human oversight, control, auditing, explainability, interpretability, solidarity, inclusion, social cohesion, science-policy link, legislative framework, legal status of AI systems, responsible/intensified research funding, public awareness, education about AI and its risks, future of employment, dual-use problem, military, AI arms race, field-specific deliberations (health, military, mobility, etc.), human autonomy, diversity in the field of AI, certification for AI products, cultural differences in the ethically aligned design of AI systems, protection of whistleblowers, and hidden costs (labeling, clickwork, content moderation, energy, resources). Several prominent companies that have established principles for artificial intelligence (AI) include Open AI, Google, Microsoft, DeepMind, and Facebook.

Some of the AI principles of open AI are

The primary objective of AI development should be to generate benefits that are widely distributed, ensuring that the outcomes serve the public at large rather than solely benefiting a select group of corporate stakeholders. The concern arises from the increasing displacement of individuals from employment due to the proliferation of artificial intelligence, resulting in a diminishing number of beneficiaries.

The underlying rationale behind the concept of long-term safety in the context of artificial intelligence revolves around the imperative of preventing any harm caused by AI systems and ensures that they do not perceive humans as obstacles or adversaries. When a company's objective is to develop an artificial intelligence system that surpasses human intelligence, this goal holds significant importance. The desire to assume a leadership role in the field of artificial intelligence (AI) development is a predictable aspiration within the realm of technical leadership. Cooperative orientation refers to a company's willingness to engage in collaborative efforts with other organizations.

Google AI principles (Fig. 3.1)

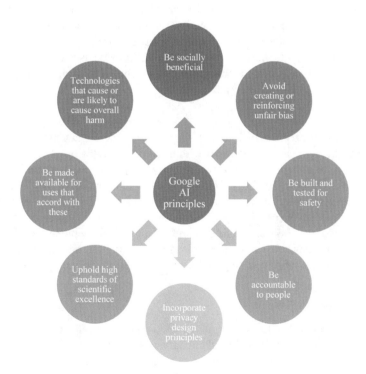

Fig. 3.1 Google AI principles (*Source* Author)

Google aims to contribute to society by focusing on various business sectors that have the potential to bring about social benefits. These sectors encompass health care, security, energy, transportation, manufacturing, and entertainment. It is imperative to refrain from generating or perpetuating unjust bias, as issues such as flawed data can result in biased artificial intelligence. The individuals in question are actively monitoring and considering factors such as race, ethnicity, gender, nationality, income, sexual orientation, ability, as well as political or religious beliefs. The construction and evaluation of a product should prioritize safety considerations, aiming to mitigate any unforeseen negative outcomes through thoughtful design.

It is imperative to demonstrate accountability towards individuals—enable users to provide feedback, receive pertinent explanations, and express their appeal towards products. The integration of privacy design principles is essential in order to ensure the provision of adequate transparency and control mechanisms pertaining to the utilization of data. Maintaining elevated levels of scientific rigor—a notable challenge associated with machine learning pertains to the considerable difficulty in replicating outcomes.

The objective is to ensure that the applications remain aligned with their intended purpose, thereby enabling their utilization in a manner consistent with the underlying principles. AI applications that we shall abstain from pursuing this aspect was intriguing as it distinguished this particular company as the sole entity that established restricted zones. Technologies that have the potential to cause harm or are anticipated to do so should be evaluated based on a cost–benefit analysis, where the advantages must outweigh the associated risks. The use of weapons or any other forms of technology with the intent to cause harm to individuals is strictly prohibited. The present discourse concerns technologies employed for the purpose of gathering or utilizing information in the context of surveillance.

Microsoft AI principles (Fig. 3.2).

Equity is a fundamental principle that should guide the behavior of AI systems, ensuring that they treat all individuals with impartiality and without bias. The principle of inclusiveness posits that artificial intelligence (AI) systems should possess the capacity to empower and engage individuals from all walks of life. The performance of AI systems should be characterized by reliability and safety. Transparency is a crucial aspect in the development and deployment of AI systems, as it pertains to the necessity for these systems to be comprehensible. The imperative for AI systems lies in ensuring robust security measures and upholding the principles of privacy. Algorithmic accountability is a crucial aspect that should be incorporated into AI systems.

DeepMind themes (Fig. 3.3)

The social purpose refers to the pursuit of objectives that are beneficial to society and are consistently subject to meaningful human oversight. The preservation of privacy, transparency, and fairness is of utmost importance, as it involves safeguarding individuals' privacy and ensuring their comprehension of the manner in which their data are utilized. The issue of AI morality and values revolves around the challenge of

Fig. 3.2 Microsoft AI
principles (*Source* Author)

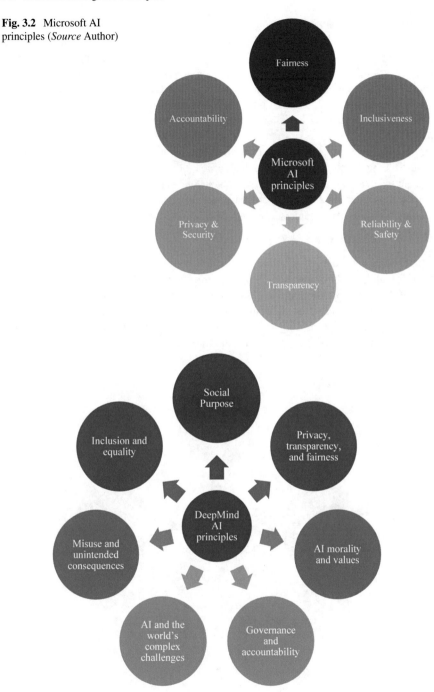

Fig. 3.3 DeepMind AI principles (*Source* Author)

reconciling divergent value systems, which poses a significant obstacle to establishing universally accepted ethical principles. Similarly, the act of endorsing values that are upheld by a majority can potentially result in the marginalization or unfair treatment of minority groups. The utilization of governance and accountability necessitates the establishment of novel standards or institutions to effectively supervise its application among individuals, states, and the private sector.

Artificial intelligence (AI) aims to effectively identify previously undiscovered patterns within intricate datasets, thereby addressing the world's multifaceted challenges. The prevention of misuse and unintended consequences is crucial in order to ensure that products are not utilized in unethical or detrimental manners. The concern regarding inclusion and equality revolves around the potential ramifications of job displacement and its disproportionate impact on certain segments of the population, leading to significant changes in economies.

Facebook AI values (Fig. 3.4)

Openness is a fundamental principle that advocates for the publication and open-sourcing of AI technologies, enabling the wider community to access, comprehend, and further develop upon them. Collaboration entails the exchange of knowledge with both internal and external partners, fostering a range of diverse perspectives and addressing various needs. Excellence entails directing our attention towards projects that we deem to possess the greatest potential for generating positive outcomes for

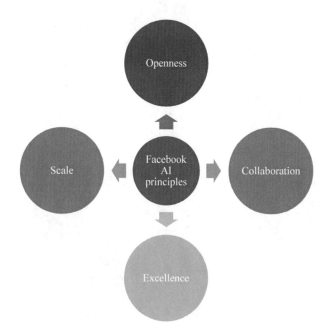

Fig. 3.4 Facebook AI principles (*Source* Author)

individuals and society as a whole. The inclusion of large-scale data and computation requirements is imperative for the development of products.

One issue pertains to the disclosure of AI principles, as it leads to increased scrutiny and examination by individuals. If there is a discrepancy between one's principles and actions, it will become evident to others. The act of a single individual disclosing confidential information can have significant consequences, or alternatively, implementing stringent measures to restrict access may inadvertently trigger the Streisand effect. The potential for rapid decline in the success of your organization is contingent upon the lack of trust that individuals place in your artificial intelligence technology. The level of skepticism towards artificial intelligence (AI) is already substantial; thus, a mere hint of uncertainty has the potential to unsettle customers.

So how do you implement principles?

The crux of the matter lies in ensuring compliance with the established guidelines. Inquiring about the potential influence of guidelines on human decision-making within the realm of AI and machine learning, Hagnedorff [15] research explores this matter. In response to the query, it is generally observed that the answer is negative, as substantiated by the trustworthiness exhibited by companies.

How can one effectively encourage a company to adhere to its guiding principles? Each company possesses unique characteristics and attributes. We are specifically addressing the impact of organizational politics on task completion. What strategies can be employed to incorporate additional directives that have an impact on the organization's financial performance?

Each company exhibits unique characteristics and attributes. One initial step involves ensuring that the model metrics are firmly incorporated. Developers typically prioritize the attainment of accuracy in their models. Regardless of the principles formulated by individuals or the company, it is imperative that during the weekly meetings focused on the development of models, the extent to which a model adheres to the principles is considered as a key metric alongside accuracy. This consideration should play a significant role in determining the model to be employed. There is a requirement for a comprehensive examination of the topic of artificial intelligence (AI) guidelines established by technology companies. This examination aims to facilitate the development of machine learning (ML) applications that are deemed acceptable within the organization and to the intended recipients of these applications.

Final words

We as product designers need to focus on the application of machine learning and leverage the potential to race through an efficient, productive, feasible, faster, and user-centered product design, development, and deployment process saving time and cost without comprising on the quality of user experience in the product and processes.

References

1. Wicaksono, I., P. G. Hwang, S. Droubi, F. X. Wu, A. N. Serio, W. Yan, and J. A. Paradiso. 2022. 3DKnITS: Three-dimensional digital knitting of intelligent textile sensor for activity recognition and biomechanical monitoring. IEEE in Medicine and Biology Society. https://www.media.mit.edu/publications/3dknits-three-dimensional-digital-knitting-of-intelligent-textile-sensor-for-activity-recognition-and-biomechanical-monitoring/
2. Ali, S., N. E. Devasia, and C. Breazeal. (2022). Escape! Bot: Social robots as creative problem-solving partners. In *Creativity and cognition*, 275–283. https://www.media.mit.edu/projects/creativity-robots/overview/
3. AHs '20., *Proceedings of the Augmented Humans International Conference*, March 2020, Article No. 23, pp. 1–12. https://doi.org/10.1145/3384657.3384799. https://www.media.mit.edu/projects/wearable-reasoner/overview/
4. Columbus, L. 2020. 10 ways AI is improving new product development. https://www.forbes.com/sites/louiscolumbus/2020/07/09/10-ways-ai-is-improving-new-product-development/?sh=230aa3995d3c
5. Iyer, H. 2022. How AI and machine learning accelerate product development workflows in manufacturing. https://blogs.nvidia.com/blog/2022/03/11/ai-manufacturing-product-design/
6. Covercon. 2021. Top 8 AI applications in product design & product development. https://convercon.com/top-8-ai-applications-in-product-design-product-development/
7. Norman, L. 2021. How to find new songs on Spotify using machine learning. https://medium.com/geekculture/how-to-find-new-songs-on-spotify-using-machine-learning-d99bd8855a18
8. Hughes, J. 2022. How the Twitter algorithm works. https://blog.hootsuite.com/twitter-algorithm/
9. Google PAIR, Research. https://pair.withgoogle.com/research/
10. JavaTpoint. Bayesian Belief Network in artificial intelligence. https://www.javatpoint.com/bayesian-belief-network-in-artificial-intelligence
11. Wodinsky, S. 2018. Facebook is making its navigation bar a little more personal. https://www.theverge.com/2018/7/31/17635758/facebook-navigation-bar-redesign-personal
12. FullStory. You can't get all the answers without all the data. https://www.fullstory.com/platform/data-capture/
13. Google Design. AI and design: Putting people first. https://design.google/library/ai-design-roundtable-discussion/
14. Jana, R., and M. Pushkarna., Six AI terms UXers should know. https://design.google/library/six-ai-terms/
15. Hagendorff, H., Ethics of AI ethics: An evaluation of guidelines. https://arxiv.org/pdf/1903.03425.pdf

Chapter 4
AI in Architecture

Architects Do Not Like AI. Is It?

4.1 AI Disrupting Architecture

AI wave versus architecture

It has been speculated that AI text-to-image software such as Midjourney [1], DALL-E [2], and Stable Diffusion [3] could revolutionize the way architects approach the conceptualization and design of new buildings and goods. In the past year, many tech firms have created programs that use neural networks, a type of AI system, to transform user-entered text into computer-generated visuals. The photos produced by these bots have become an Internet sensation, with many people wondering what effect they will have on the future of design and architecture because of the uncanny realism of the imagined buildings they produce. Bill Cusick has worked with the software of the well-known visualization firm Midjourney, and he is the creative director of one such company, Stability AI, which has developed text-to-image software called Stable Diffusion and DreamStudio. It is the basis for the future of creativity, which is why the software is so important. There is a type of playfulness in design that has hitherto been unseen in architectural design and representations [5].

Even among designers who are well-versed in the technology, there is much discussion about where it will go from here. "It is purpose is to rapidly capture the spirit of a project." Matys Design's Andrew Kudless has speculated that the technology would soon replace traditional brainstorming and sketching in the earliest stages of any given project. He went on to compare AI imagery to architectural sketches in that it is frequently colorful and dreamy but not always actionable as a precise plan. Whenever Kudless is sketching with a pencil, he doesn't stress over whether the column or whatever is in the exact perfect location [5]. For the simple reason that is not the point of a sketch. This is not an attempt at precision or extreme accuracy. Its purpose is to depict the desired outcome of a project swiftly and accurately. Stable Diffusion can be used for the architectural sketching phase. Bhatia specializes in conceptual work depicting futuristic cityscapes that feature intricate fusions of manmade and natural elements.

Artificial intelligence might be used to speed up the time it takes to go from having an idea to putting it into action. Currently, AI is being used by designers and other creative professionals mostly for ideation, but this is expected to change as AI becomes more integrated into the design process. Using AI, we can shorten the time it takes to go from "idea to execution." Kudless, a trained architect, currently works in Midjourney, where he constructs evocative structures (often draped in fabric) that, in his opinion, demonstrate how material and scenery not typically included in initial architectural plans can change the way in which projects are presented to clients [5]. To this day, architectural firms frequently win commissions based on conceptual drawings that are anticipated to be a near-exact representation of the finished structure. We should start providing renderings to clients nearly from the beginning, he suggested, to help them adjust to the idea that the design would evolve over time. That's great for the client-designer dynamic because it gets the client involved in the process in a significant way, as one of us put it.

Architects and designers alike agreed that the program may revolutionize their industries' standard methods of getting work done. Kudless remarked, "It is really aggravating at times the amount of labor that is being demanded of especially younger employees within firms to produce render graphics." It is ok if we can automate part of that while maintaining control [5]. Designs that take simple patterns and repeat them in different circumstances, such as repeats for offices and parking lots, could be useful in enhancing the design process. The technology had enormous promise for liberating labor during the early stages of design because it generates visuals by creatively combining well-established aesthetic and formal elements. It has become about the beauty of merging forms that are utterly opposed. It has become about taking diverse styles and combining them.

On the other hand, Kudless thinks that text-to-image software is more of a useful tool than a replacement for architects since it still takes knowledge and ability on the part of the user to achieve desirable results [5]. Natural selection is not unguided like this. Designers may generate many images when using the tools, which they must subsequently manually modify. Interior design could be the next step for AI images.

The AI systems recognize the text entered to create an image, but they do not comprehend language in the same way that people do, often misinterpreting context or focusing on secondary nouns and adjectives. Therefore, finesse is required while selecting the appropriate words to produce something of value. Artist Bhatia, who makes conceptual works depicting futuristic cityscapes that display complex integrations of architecture and environment, noted that his software is a tool that needs a human equivalent to function [5]. Images created by AI are reliant on textual input representing the designer's intent, which can and will vary from user to user. Cusick compared the software to the game of chess, saying that both may be learned quickly but taken decades to perfect. A dynamic user is something it just cannot perceive.

AI-driven design goes well beyond the built environment. Interior design solutions on Cusick's company's platform, he says, have shown a lot of promise. Industrial design has also advanced, thanks to text-to-image converters. American designer Dan Harden of Whipsaw, an industrial design firm, has noted that artificial intelligence is making its way into the field gradually, albeit in a basic form [5]. Digital design is different from product hardware design since AI is incorporated into a common software platform.

The complex interplay between the three-dimensional aspects of a product, its interaction with users, and its environment presents significant challenges for artificial intelligence (AI). This is primarily due to AI's limited ability to grasp the intricate nuances of human interaction within the physical realm in which a product exists.

Harden commented "It is challenging to simulate the actual design process with ADI. The concept of "embodied cognition," the idea that a living thing's physical presence affects the way it thinks, is absent. This theory holds that the mind is not just linked to the body, but that the body influences the mind." However, Harden is confident that AI design will eventually approach these embedded concepts [5]. Though he thinks AI software systems are more likely to prioritize art, animation, and film. Future breakthroughs in text-to-image technology may have increasingly obvious impacts on designers. The next big thing in software, he predicted, would be AI that can generate 3D visualizations or films in response to textual input.

There will be widespread career shifts once the technology made its way into moving pictures. Cusick claims that Stability AI is actively trying to create 3D processes for their platforms, and that the technology also has the potential to enable voice-to-image and image-to-image conversion [5]. There are tendencies to be biased. The utilization of preexisting image pools to generate new visualizations is a major source of bias in text-to-image technologies. There is an imbalance of representation between different architectural types and between real photographs and computer-generated ones. It can be compared to traditional teaching methods, claiming that favoring one architect or design style over another can lead students to develop prejudices.

It is imperative to ensure that the software receives a sufficient amount of data in order to avoid the development of a self-reinforcing feedback loop that exclusively enhances its performance in tasks explicitly assigned to it. Cusick has engaged in discussions within the field regarding the potential of utilizing image-to-image processes to address biases present in neural networks and the datasets of images and text that are utilized in these programs [5]. This feature would enable architects to incorporate diverse architectural styles by utilizing an image-based input mechanism, thereby facilitating the creation of a composite or synthetic architectural composition. The discussion revolves around the creation of cultural models capable of comprehensively capturing cultural data from any given country. Additionally, the Stability AI team is currently engaged in deliberations regarding the potential development of culture-aware neural networks.

The idea is to make people think of a future where architecture coexists with nature. Designers like Bhatia believe that technology may bridge the gap between humans and nature by displaying structures with natural qualities, like the shape of a tree, that AI can capture with great accuracy [4]. The goal is to get people thinking about a future where architecture can coexist with nature. With the current level of investment in studying how to improve construction techniques and materials, we will be able to create structures that are more in harmony with their natural surroundings. We are less optimistic about the potential of technology to realize utopian ideals. He predicted that cultural prejudices would always be present, and that utopian society was an unattainable goal. It's a mirror, and it aids in our self-awareness. The creators of Clippings, an online marketplace for finding furniture, have expressed optimism that AI will empower designers [5]. Oliver Wainwright, an architectural critic, has used DALL-E to develop a hypothetical design for the Serpentine Pavilion [6].

AI versus architects job

Will Wiles asserts that despite the justifiable debate surrounding AI art, architects should not fear about being usurped by software that can generate images of structures. However, there are still a couple of things we can count on, despite the current climate of uncertainty [7]. The first is the permanence of AI-created works of art. The second is that there is good reason to be skeptical about AI-generated artwork. Artists on Earth are understandably worried that it will take over their day jobs. A further insult is that the AI appears to have some plagiaristic tendencies.

AI is used by websites like Midjourney [1], DALL-E [2], and Stable Diffusion [3] to generate art based on user input. This strategy relies less on innate intelligence and more on training computers to jumble and unscramble images until they know how to construct an image from noise [7]. The best results still have that characteristic sludgy look, and they often struggle with subjects like the human hand. They are not quite there yet, but enough people are worried that underappreciated human artists will not be able to make a living if people start telling computers to copy the work of their favorite artists. There is an increasing perception that the development of an AI architecture is a distant prospect. In considering the field of architecture, it becomes evident that its scope extends beyond superficial observations. In the future, it is conceivable that a singular software application could possess the capability to envision a structure, generate comprehensive and feasible architectural drawings, intricately design drainage and gutter systems, and effectively engage in collaborative efforts with clients, contractors, and relevant authorities. This level of functionality would mirror the expectations typically associated with a human architect. But is there evidence that computer-generated representations of buildings are beginning to threaten architects' cherished capacity to conceptualize new building designs? Even if AI-generated ideas are never put into construction, they can nevertheless have a significant impact through paper architecture.

The gowned players in Théâtre D'opéra Spatial, Allen's award-winning AI-generated image, stand in front of a large, cavernous interior dominated by a gigantic circular aperture. More and more individuals are turning to various mediums to produce similar architectural renderings. There are over 32,000 posts on Instagram with the hashtag "#Midjourneyarchitecture," and the Instagram account Midjourney Architecture, which compiles fascinating examples, has over 34,000 followers [7]. Most of the "architecture" displayed is completely made up, although there are a few instances where it looks hauntingly realistic. An ever-increasing user base is turning to various mediums to produce architectural renderings.

Midjourney construction is often persuasive enough to cause waves in the physical world. Hassan Ragab, an Egyptian artist, and designer living in California, has been utilizing Midjourney to try out new concepts and techniques, as well as to make hyperbolic homages to the Islamic architecture of ancient Egyptian cities like Cairo and Alexandria. An astounding 8.5 million Instagram users viewed one of his "Cairo designs" depicting an Escher-like wall of windows joined by complicated and impossible intertwined brickwork and embellished with Islamic tracery.

Artificial intelligence architecture complements these areas, although it yields less reliable outcomes. Allen's architectural setting is spectacular at first glance, yet there is more going on than meets the eye. It is merely a representation of baroque flourishes and refined architecture, rather than the real thing. It displays the peculiarities typical of AI architecture, such as a smeared, asymmetrical design with fudged, impossible edges, joints, and details. We are still living in the world of images rather than reality [7]. The machine manipulates two-dimensional pixels, not physical objects, or physical space. Yet, AI is not threatening to replace architects, but it is not hard to imagine computer-generated graphics beating out human artists in painting or drawing competitions. The practical application of artificial intelligence is more probable in domains where computers are already extensively utilized, such as the creation of residential floor plans or the resolution of service layouts, rather than in innovative sectors that require imaginative thinking. The observation of recurring resonances between the architectural works of several practitioners with highly distinctive styles, such as Gaudi's biomorphism, Eisenman's uncanny geometries, and Hadid's fluidity, is indeed intriguing. Artificial intelligence lacks the capacity to discern and question established norms. It is advisable to discourage hardline ideologues and style partisans in the field of architecture who rigidly adhere to objective notions of beauty and perceive any deviation as both a moral and aesthetic deficiency, due to the potential negative consequences it may generate. Contrary to what the machine thinks. Degeneracy is baked into the very fabric of generative architecture, making it both unique and intriguing [7].

AI versus architects

Potentially, architectural services will not be needed in the foreseeable future. As AI develops, it will soon be able to independently create a building's blueprints. AI can change the architecture profession by creating designs faster and more accurately than ever before, displacing the need for human architects. Concerns about the future of architecture in a world where buildings are increasingly being created by artificial intelligence have been raised considering this development, which could mark the end of the profession as we know it. In case you were wondering, I did not pen the preceding text. ChatGPT, a state-of-the-art artificial intelligence text generator, is responsible for its creation. Don't be fooled by its name; ChatGPT is much more than just a chatbot. The foundation of this system is GPT3, a huge Generative Pre-trained Transformer (GPT) that use deep learning to generate natural-sounding text in response to user input [8].

The architectural community is becoming cognizant of the vast opportunities presented by AI. This is largely attributable to the exceptional image-generating power of GPT3-based "diffusion models" as DALL-E, MidJourney, and Stable Diffusion. Occasionally, the created photos have a quality that is nothing short of mind-blowing. Despite their incredibleness, these photos could be a trap. Some architects' fixations with them have caused them to miss the forest for the trees. In the end, AI's contribution to the design process is what makes it revolutionary, not image generation [8]. Some occupations are already in jeopardy due to ChatGPT, though perhaps not the ones you'd expect. We can easily picture Amazon warehouses and Tesla assembly lines devoid of personnel, with the blue-collar workers being the first to be laid off. Unfortunately, robotics development has been sluggish. There is still a long way to go before a robotic arm can accomplish anything as basic as choosing and picking up a brick. Meanwhile, advances in AI have accelerated to the point that ChatGPT is fully capable of coding. Only the human mind, with its extraordinary capacity for original thought, can ensure its continued success. Still, the future of the architectural profession is in jeopardy, but there are already warning signals that AI is becoming not just good but terrifyingly good. AI functions and will function as a prosthesis, enhancing and extending the capabilities of the architect by adding new features and capabilities.

Have a look at Hassan Ragab's Cairo Sketches. Take comfort, architects: artificial intelligence isn't going to steal your profession just yet [7]. Obviously, this can be of great use. With the help of AI, even a one-person office can take on larger competitors. The implication, however, is that fewer architects will be needed by private firms. According to Xkool's CEO, Wanyu He, a single architect utilizing AI may accomplish as much as five architects without AI. Is it safe to assume that 80% of the profession of architect is currently in jeopardy? Eventually, it seems, AI will be able to come up with building plans on its own, rendering architects obsolete. The issue is not with AI per se, but with what it can do. Eventually, it will surpass humanity in every conceivable way. After his historic loss to IBM's Deep Blue computer in 1997, chess grandmaster Garry Kasparov said, "Everything we know how to do, robots will do better." The time to talk about answers is now. If architects want to succeed in today's cutthroat industry, they must embrace AI; as the adage goes, "if you can't beat 'em, join 'em." We must become "superusers," a concept developed by architect Randy Deutsch, by becoming acquainted with AI's capabilities and subsequently upgrading ourselves [8].

Instead of developing yet another structure, architects should focus on designing the future of their industry. The first step in solving any problem is realizing there is one. For once an issue is identified, it is no longer a problem that traps us but one that we can solve. Surely, rather than developing yet another structure, architects today should be working on the future of their profession. AI has the potential to enhance architectural design by providing tools for automating repetitive tasks, analyzing complex datasets, optimizing designs for energy efficiency and sustainability, and generating new design ideas. AI can also assist in the decision-making process by simulating various scenarios and predicting their outcomes, helping architects make more informed choices. However, AI should be used in conjunction with human creativity and expertise, rather than replacing it, to ensure that designs are aesthetically pleasing, functional, and responsive to human needs.

4.2 Design Tools on the Foundation of ML

Rethinking design tools

A person's ability to think creatively is growing. Designing is becoming increasingly accessible, as the variety of tools, platforms, and devices available increases as their prices decrease. You can create your own movie, album, city, or even print your own flowerpot. All of this is convenient to do on a personal computer or even a mobile device. Naturally, we are all eager to give these novel opportunities a try. We seek the liberty to create for its own sake. It's appealing to imagine ourselves as modern-day Renaissance men and women [9]. The tools may be more accessible and less expensive, but that doesn't mean it's easy to make an impactful image or tell an interesting tale. It is as challenging as ever to come up with a truly novel statement. It remains as challenging as ever to organize the various components that make up an aesthetic experience. Expertise, practice, and experience are still required for these tasks. Tools for creating designs and programming languages have given us a great deal of freedom. However, until we know how to put that power to good use, it will never be truly ours. We should not merely employ it but also develop our own unique approach to the medium. The purpose of design software is not to make you feel creative. What you end up with should be something that you can call your own with their assistance.

Design tools and machine learning

Machine learning can help us to simplify design tools without limiting their expressivity, without taking creative control away from the designer. This may seem totally counter-intuitive. When we think of machine learning or artificial intelligence, we think of automation. Creating a unique design always requires several choices. It takes time to make a choice. Because of this, design tools have gravitated towards two poles: The one-size-fits-all approach is typical of consumer-level design software, which attempts to streamline the design process by confining users to a small set of standard templates. In contrast, professional design tools sometimes take a "kitchen sink" approach, providing an overwhelming number of low-level capabilities that have steep learning curves and often don't align with the user's style of thinking. To the untrained eye, machine learning appears to be a little more nuanced "one-size-fits-all" approach, simplifying design processes by relieving the designer of some decision-making tasks [9]. This is a possible and even plausible application of machine learning, especially in its infancy. On the other hand, it presents many more varied opportunities. While it is true that we cannot reduce the sheer volume of choices we have to make during the design process, we can alter the factors that go into those selections.

Let us discuss how machine learning can alter our experience with and reliance on design software. Emergent feature sets, descriptive design, process organization, conversational interfaces, and exploratory design are all examples. These concepts could improve efficiency in the design process without interfering with the designer's autonomy. But what is even more fascinating is that these processes will free designers from the burden of figuring out how to adapt their work to the constraints imposed by every given tool. That is to say, the tool will follow the designer's instructions, rather than the other way around.

The emergence of characteristic groups

When a designer gets down to create a design, they may or may not have a clear mental image of what the final product will look like. Either way, they must figure out how to get there by discovering the steps that must be taken with various tools to turn a blank slate into the finished output.

> It is the job of the sculptor to unearth the statue that is hidden within each block of stone— Michelangelo

This remark resonates with me because it portrays the creative and design process as a quest. There are an endless number of conceivable sculptures that could be carved out of a block of marble within its finite confines. The role of the creative professional is to unearth the "needle in the haystack," or the unique mix of qualities that fulfills a specified demand. This is akin to the quest of a chemist for a new chemical or the quest of a chef for a novel flavor profile. While each design challenge will have a unique combination of interconnected attributes and constraints, and their corresponding search-spaces will be unique as well, the underlying method will be similar. Let us look at what would go into the design of a commonplace item like a glass. If the glass is to be made taller, its base will need to be widened to prevent it from toppling over. In this case, we're adjusting two attributes relative to a single limitation. At what point does this property ratio cause the glass to tip over? Experimenting inside the search-space allows us to hone our skills, as we discover the interplay between various properties and the underlying physical restrictions that govern them.

Let's visualize this search-space as an enormous map, with individual coordinates for each probable destination. Each function in the program is represented here by a road that leads us some distance in some direction. Compared to a major highway, a low-level feature (within a professional design program) would be more akin to a local street, taking us only a short distance. A series of low-level commands could be used to expand the glass's volume. We may, alternatively, condense this sequence into a single, high-level feature, like those seen in consumer-level design software. In practical terms, this element would function more like a thoroughfare. One of the amazing things about highways is that they allow us to travel long distances with a minimum of effort. However, exit ramps from highways only tend to be located at heavily traveled hotspots. Taking local roads to more out-of-the-way locations increases the number of individual steps required to complete the journey. However, we are not even given the option to use local roads in most consumer design tools. Maybe if we had gone a little further (around the distance of the exit ramp), we would have found a better place to go, but now we're stuck. Moreover, it is possible that we are unaware of this potential alternative outcome and the impact it would have on our broader design objectives. While high-level tools can help us navigate across the search-space more rapidly, they often come at the expense of limiting our capacity to be as expressive or as flexible as possible. As a result, most people will stick to the most convenient areas, leaving the rest of the map untouched.

A person that is set on reaching their goal may encounter several obstacles along the way. However, it is quite improbable that the user will get there naturally if the specific qualities of that destination are not thoroughly specified in his or her mind in advance. A high-level tool may only render inaccessible a small portion of the map, but due to its fragmentation of the search-space, much of the map is rendered inaccessible. When it comes to expanding our creative horizons, consumer-level design tools really limit us. To keep our options open, it seems we must either limit ourselves to low-level operations or produce an overwhelming number of high-level capabilities that would accommodate a larger variety of use cases at the expense of the brevity of the tool's vocabulary. Ideally, every time we set foot outside our front door, a brand-new highway would be built for us, connecting us to any place on the planet in a matter of minutes. However, this can't be done using only the built-in high-level features that are now available. Thanks to machine learning, we can infer a wealth of information about users and their goals from their actions alone. Making tools that learn from the designer's interaction with the software is preferable to trying to predict their needs through a prebuilt, high-level feature set.

Conceived via inquiry-based design

An appreciation for what is aesthetically attractive or practically beneficial is some-thing everyone is born with. However, many of us lack the lexicon, methodology, or self-assurance to apply these intuitions towards actual creative output in a field of design in which we have no prior experience. Tools for designers should not just aid in the execution of designs in areas of expertise, but also in the acquisition of knowledge in unexplored areas. A lot of individuals would be at a loss as to where to begin if we stopped them on the street, handed them a sheet of paper, and asked them to draw their perfect living space.

Many individuals would find the work far less daunting if we gave them access to Pinterest and asked them to design a living room by picking and choosing things they like. There is no need for the user to try to remember the function's esoteric name within a convoluted menu structure and the specific actions it performs. While an alternative, people can watch as the behavior is applied to a duplicate of the scenario and then decide if they like it or not. Earlier, we had a look at a 2D representation of a 2D search-space. This representation is not comprehensive, but it does provide a simple and fine-grained framework for design. Any design feasible within the constraints of the search-space can be reached with just a point from the user.

This spatial organization aids the user in forming a distinct mental model of the impact of a given translation within the search-space. Obviously, in the actual world, design issues rarely have only two dimensions of variability. However, a low-dimensional map of a high-dimensional feature-space can be generated using a machine learning technique known as "dimensionality reduction."

Let us pretend, for the sake of argument, that we have already drawn an oak leaf above using Bezier pathways, only to realize that we want it to appear more like a maple leaf. This transformation must be mapped to the logic of a Bezier route before it can be implemented in conventional design tools. This process is second nature to us because of our profession as designers. However, the skill to control a Bezier route is only remotely relevant to the issue at hand. There's a good chance that if we try to convert one shape into another, we will end up undoing a lot of our hard work. Because of this, the price of discovery rises significantly. It is unacceptable that things are this way. We should not have to go halfway across town just to visit our next-door neighbor. Instead of being constrained by the tendencies of a single tool's approach of abstracting the path to a given destination, designers can follow their own design intuitions with the help of a variation control surface. By taking this approach, we are not limiting the designer's agency; rather, we are liberating them from the supplementary requirements and conceptual remapping of older tools. This allows them to concentrate on mastering the design choices themselves rather than the technological means through which they are implemented.

A step-by-step description-based design process

We have studied interface types that are extremely like maps in the traditional sense so far. We can use textual labels, or "street signs," just like on any other map. With spoken directions like "transport me to a maple leaf," we can now easily move around the creative environment. When we get there, we can ask to be brought "a little closer to an oak leaf." All by itself, this is very effective. But it turns out there is more we can do with this concept.

Low-dimensional word-maps representing conceptual linkages were first described in a series of works by Tomas Mikolov and colleagues published in 2013 [9]. Words that are semantically related in everyday conversation will be close together in the word embedding map, just as they are in the previous maps we've been considering. Even more astonishingly, Mikolov and his coworkers found that word vectors can be subjected to algebraic changes that have conceptual significance. In other words, they found that algebra may be applied to concrete ideas. As an illustration, they demonstrated that the outcome of the phrase "word vector" is That is right, "Madrid, Spain and France" is the word vector closest to "Paris." besides the fact that King minus Man plus Woman yields a vector near to Queen. This intriguing mechanism suggests a fresh approach to the use of design software. Without resorting to external abstractions and control mechanisms, we may now perform visual operations on language notions. If, for instance, we wanted to find an aesthetic reminiscent of Picasso's work but not from his most mature (analytical) Cubist period.

The same applies to audio or any other form of data. Recent years have seen the development of similar methods, such as "Style Transfer" and "Neural Doodle," which further expand these mechanisms. These methods have been incorporated into photo-sharing apps, however not as components of comprehensive design tools but rather as a fresh visual effect like to those seen in photo editing programs like Instagram and Photoshop. A novelty display of this functionality quickly becomes kitsch and accomplishes little to re-conceptualize or extend design processes, as the Photoshop "filter bubble" of the 1990s demonstrated.

However, when combined into a bigger and more holistic design framework, these methods offer a potent mechanism for manipulating media inside their original vocabulary, without mapping them onto an abstraction. As a result, we may directly manipulate the idea space in which ideas exist, allowing us to do things like explore that space and create new ideas. Nevertheless, I believe there is something missing, no matter how revolutionary the techniques may be. The hardest part of design, as any designer will tell you, isn't the process of making decisions. It's challenging to create a unified whole from a variety of separate selections. As designers, we constantly switch our focus from one component decision to another while keeping the big picture in mind. It's possible for these separate choices to come into direct opposition with one another.

We cannot finish one side of the puzzle like a Rubik's cube and move on to the next. Because of this, we would have to undo some of our previous efforts. We need to work on all the issues at once. This is often a convoluted procedure, and mastery of it is important to the designer's craft. These component decisions can be simplified with the help of the machine learning approaches we have examined still they do not completely solve the design problem's biggest challenge. Let us dive deeper into two additional ideas that will assist designers get this kind of knowledge.

Interaction design for processes and conversational systems

Compared to complicated, multifaceted statements, simple expressions that represent a single instruction or point of information are more easily understood by machine learning algorithms. Nonetheless, decomposing a dynamic, complicated system into its component elements is one of the trickiest aspects of any design process. Design tools that aid the designer in this step would be helpful. Tools could aid the designer in providing brief assertions by facilitating the development of interfaces and workflows that guide the user through a set of relatively straightforward steps designed to help them accomplish a bigger, more involved goal. 20Q, a digital take on the game twenty questions, is a great illustration of this method in action.

20Q is a variation on the classic car game that has the player think about an item or well-known person and then asks them a series of multiple-choice questions to figure out what they are thinking about. In this procedure, the initial inquiry is always whether the item in issue belongs to the animal, vegetable, mineral, or concept categories. Follow-up queries find out if there are any further differences you can make based on the user's input. A follow-up question might be "Is it a mammal?" if "animal" was the initial guess. A follow-up inquiry might be "Is it normally green?" if "vegetable" were the initial response. To answer the following questions, select yes, no, sometimes, or irrelevant. When asked twenty questions about a person, place, or object, 20Q is 80% accurate, and after twenty-five questions, it is 98% accurate. It employs a machine learning method called a learning decision tree to figure out the shortest path to the right answer by asking the right questions in the right order [9].

The algorithm learns the importance of each question by analyzing the data provided by past users' interactions with the system and displays the questions with the highest importance to the user first. If it were established that the user was thinking of a famous person, the next question should probably be whether that person is still alive rather than whether that person has written a book, since few historical figures are still with us today, but many famous people have written books of one kind or another. Although none of these questions by themselves capture the whole scope of what the user has in mind, a few well-chosen questions can reveal the proper answer with surprisingly little effort. This approach not only helps the system understand the user's expressions, but it can also help the user express themselves more precisely and effectively.

This procedure can be broken down into a series of smaller steps that each serve to uncover the best possible course of action among a wide variety of interconnected choices. A translation vector across concept-space, each question-and-answer interaction brings the user closer to the desired output while simultaneously encouraging them to consider and communicate the idea in its entirety. With this approach applied to a design interface, a user can zero in on a specific look by answering a few questions. These inquiries could be responded to vocally or by more natural means of communication such. Recent developments in machine learning have made it possible for machines to provide answers to more sophisticated questions posed by users regarding the qualities of a design in a variety of contexts.

For instance, the user might ask specific questions to better assess the design's functionality. Such a conversation would mimic human speech patterns while taking advantage of the machine's exhaustive understanding of the design's characteristics. This may be related to the machine's capability to simulate material, physical, or chemical limitations. An architect, for instance, may save a lot of time by incorporating this skill into a real-time interaction and promptly discarding concepts that are not likely to pan out. In addition to the aforementioned "real-world constraints," it is possible that the user's intended meaning in each interaction is not always evident, either due to the limitations of the machine's knowledge or the user's lack of clarity in their own statement. Instead of taking a "best guess," the machine might pose clarifying questions and provide alternate solutions. By conversing with the machine, the user's intent can be better understood, and the machine's knowledge can be expanded. Also, unlike an "Action History," the user's iterative process may be easily retrieved and reflected upon with a conversational approach.

By unfurling the interface into a linear, explorable "newsfeed," the user can examine each step in his or her thought process and simply return to earlier iterations, branching off in a new path while keeping retaining each other version of the design.

Conclusion

Many individuals appear to be concerned that AI will eliminate our need for human labor. For our part, I like to imagine an alternate, more hopeful future in which we continue to play a significant role. In this potential future, we are all stronger and more human than ever before. It's not that we're going up against inanimate objects in the future; rather, we're just exploiting them to further our own goals. On the way, though, we'll need to refresh our memories about the functions of various implements. Devices are not intended to lessen our workload. Really, it doesn't. They are supposed to give us an upper hand, allowing us to exert more pressure. To move rocks, tools are used. Cathedrals are constructed by human hands.

4.3 Personality in AI

The significance of developing a personality for an artificial intelligence system. How to develop an artificial intelligence personality?

A personality refers to the unique characteristics, demeanor, and approach through which an application expresses itself in communication. The term "it" refers to a particular entity, which is characterized by a collection of attributes that determine its visual, auditory, and tactile qualities. The appropriate linguistic choices and tone that encapsulate your application and distinguish it from rival offerings. It is highly probable that your application and company possess an inherent personality. The existing web or app design inherently embodies the corporate identity of the company. The brand is comprised of various elements, including color selection, typeface choices, user interface layout, documentation, and error handling. Essentially, the brand is determined by the corporate identity. The subsequent phase entails leveraging the personality previously employed for the brand and applying it to train the artificial intelligence system. Currently, there exist certain companies that lack a distinct corporate identity. One possible explanation for the lack of a clearly defined personality in many companies is the utilization of preexisting templates for their websites or applications. Numerous templates and default frameworks are available for the construction of website widgets and application interfaces.

Currently, there is no existing template available in the field of artificial intelligence for the specific task mentioned. The process of developing a personality necessitates an individualized approach in each unique circumstance. The Clippy avatar endeavored to maintain a lighthearted atmosphere by incorporating humorous anecdotes alongside its assistance. The issue stemmed from the corporate branding of Microsoft Word, which elicited frustration due to its unforeseen behavior. The presence of jokes or unconventional interface elements can potentially enhance users' interest and motivation to explore an application, but only if these elements align with their expectations. The selling power of personality is significant, thus making it a worthwhile investment if executed effectively. The perception of a company's enthusiasm and passion can be discerned by individuals. In the near future, there will be a notable shift in which artificial intelligence will become conspicuously apparent due to its lack of distinctiveness. Adhering to optimal methodologies will result in achieving a position of average performance. The issue at hand is that a significant number of users express dissatisfaction with their application, characterizing it as "not terrible." There are already companies that are focused on the development and implementation of artificial intelligence. Google is currently seeking individuals with creative abilities to enhance the incorporation of humor and storytelling into interactions between humans and machines. Similarly, Microsoft Cortana's writing team consists of professionals from diverse literary backgrounds, including a poet, a novelist, a playwright, and a former television writer. The development of personality traits can be facilitated by individuals in various fields, such as writers, designers, actors, comedians, playwrights, psychologists, and novelists. The job descriptions typically associated with technology companies are not what one would consider conventional. The incorporation of these proficiencies into technological positions has given rise to designations such as conversation designer, persona developer, and AI interaction designer [10].

Now that the necessity has been established, let us proceed to discuss the process of creating a personality. If one desires to engage in long-term planning, the following predictions are offered. In the foreseeable future, it is likely that companies will offer a diverse range of personalities to enable individuals to select their desired voice or physical appearance based on the artificial intelligence user interface. Various design styles, such as Google's material design and Microsoft's metro, can gain popularity among individuals with different personalities. The future prospect of templatizing a personality, similar to the existence of WordPress templates, suggests that it is inevitable that companies will engage in legal disputes over the infringement of AI personalities, akin to contemporary brand infringements. In my personal opinion, I am eagerly anticipating the future when a sufficient amount of user experience (UX) research has been conducted to determine the most effective custom artificial intelligence (AI) personality for interaction modality. It may appear unconventional that the most reliable source of legal information is an individual conversing with a seasoned sailor. However, it is imperative to establish a specific interaction modality in order to ascertain the personality that becomes associated with it.

Is it not advisable to express one's own personality? Or the personality of the founders?

There are several issues associated with this matter. The capacity to establish a business does not invariably result in effective customer engagement due to a multitude of factors. Another factor contributing to the ineffectiveness of this approach is the inherent difficulty in self-assessing one's own personality traits. The majority of individuals tend to primarily identify themselves with their positive attributes; however, it is regrettable that they often overlook the accompanying negative traits. How can we determine the most suitable personality traits for the users? The individuals in question are queried for their input. Conduct a survey to elicit individuals' choice of adjectives for descriptive purposes. Utilizing a standardized list, such as the Microsoft Word Association Test, is deemed most effective due to the inherent requirement of allocating sufficient time to achieve equilibrium between positive and negative words, while ensuring comprehensive coverage across all domains. Word associations are often considered a convenient and expeditious approach to proceed with. However, in the event that this method does not align with your preferences, it is worth noting that some individuals have reported successful utilization of the Myers-Briggs Type Indicator (MBTI) to delineate character attributes. The claim has been discredited due to its tendency to oversimplify the categorization of personality types. However, this approach proves beneficial as it facilitates the process of decision-making by simplifying the available options. An alternative method for collecting the information is a technique known as spectrum.

The Five Factor Model was developed by Ari Zilnik [10]. Personality is defined as a composite construct encompassing the following dimensions: openness to experience, conscientiousness, extraversion, agreeableness, and neuroticism. It is important to note that responses often exhibit a bias towards positivity; thus, it is advisable to allocate greater attention to negative feedback. The purpose of this inquiry is to solicit users' selection of terms that they closely link with our brand, company, and application.

There are five distinct domains that can be assessed:

- The level of awareness exhibited by the user with regard to the company, product, and the underlying needs for the product is a key aspect to consider.
- The concept of consideration pertains to the perception of quality and value associated with a product. Individuals may experience difficulties in comprehending or locating specific features.
- Preferences play a crucial role in distinguishing products from their competitors by virtue of their distinctive features.
- Action refers to the process of accomplishing tasks and completing objectives.
- Loyalty refers to the likelihood of a user choosing to utilize an application on subsequent occasions.

The interface for the user may incorporate various aspects such as sound, haptics, visuals, or augmented reality/virtual reality (AR/VR), depending on the specific functions that the AI is designed to assist with. If any similar resources or tools are available for the present interaction, please provide feedback on those as well. Sound associations can be established through the examination of companies that have obtained patents for their distinctive sounds, such as Porsche and Harley Davidson. When engaging in conversation with users to elicit word associations, it is crucial to gain a comprehensive understanding of their personality, objectives, current life stage, and, most significantly, their aspirations. This will be relevant in subsequent stages. The subsequent phase entails conducting word association tests with internal individuals, focusing on the perspective of aspiration. What is the intended direction for the product as envisioned by the decision makers? This is ideal as it aligns with the primary concerns of project managers and stakeholders. Next, we proceed to the comparison.

The present perception of oneself versus the desired future perception

Evaluate the received responses and assess their alignment with predetermined expectations. One should not anticipate the word associations to be precisely identical, although they should not deviate excessively either. If there is a significant deviation from the customers' perception, it is necessary to engage in open and honest dialogue regarding the current state of product excellence within the company, or alternatively, to undertake extensive efforts to strengthen the foundational aspects. Excessive ambition in one's claims or assertions can engender a perception of unreliability. One's identity is determined by their inherent characteristics and personal attributes. More-over, when the reach is excessively extended, the probability of making errors signif-icantly increases. If an incorrect outcome is obtained, a substantial amount of effort will be rendered futile. Once an individual has developed a vision of how they are perceived in an idealized manner. Evaluate the objectives in relation to the customer's objectives. These should also be in close proximity. An illustrative instance would be to visit a retail establishment specializing in clothing for adolescents. The personnel exhibit a tendency to closely resemble the individuals featured in their promotional campaigns. The alignment between stores and their target aspirational demographic for customer interactions is not a mere coincidence. From a theoretical standpoint, how would the employees in your organization be characterized? In what manner do your customers demonstrate alignment with their peers? What factors serve as motivators for individuals to engage in their chosen actions or behaviors?

One aspect that necessitates clarification within the realm of personality pertains to:

A Comparison of Professional and Casual Communication Styles—When adopting a fully professional approach, it is important to avoid completely elimi-nating one's personality. It is important to note that when adopting a casual approach, the level of informality may vary across different groups and cultures. Therefore, it is crucial to gather comprehensive information about all target markets. The level and type of humor. Which type of humor do you employ: dry or silly? The choice between being a generalist or a specialist is a topic of academic interest and debate. Are you attempting to achieve a rapid and efficient conversion? Alternatively, could the entire bot encounter be intentionally designed to foster long-term engagement as a component of a broader creative campaign?

There is no necessity to present any of the word associations to the developers. When attempting to rephrase the brand guideline, the resulting language consis-tently gravitates towards terms such as innovative and progressive. When composing dialogues, it is assumed that the brand guidelines will not be readily accessible. Devel-opers typically do not have physical objects, such as them, present on their worksta-tions while they are engaged in the process of writing code. There is a requirement for increased simplicity. If one were to liken their organization to a prominent indi-vidual, which person would best embody its characteristics and values? Given the existing set of personality traits and aspirations, to whom does it pertain? The act of reducing everything to the terms "innovative" and "progressive" can be considered as an equivalent representation. It is advisable to select distinct personality traits for each of the five primary touchpoints, namely Awareness, Consideration, Preferences, Action, and Loyalty. The application may exhibit a higher degree of specialization, resulting in variations in its areas of focus depending on individual requirements.

Having established the appropriate behavior for the AI in various scenarios, it is now pertinent to discuss certain aspects that should be refrained from. Currently, conversational artificial intelligence systems have reached a level of proficiency that enables them to convincingly emulate human-like interactions. In due course, given the present state of technology, it is likely that your AI will eventually exhibit char-acteristics that place it in the less favorable region of the uncanny valley. Asserting or disavowing one's humanness when prompted can be a complex task, as it may be more convenient to abstain from making such a claim, while simultaneously avoiding outright denial. The principles governing the personality of Google Assistant discourage the termination of conversations by rejecting the notion of human-like characteristics.

The subsequent challenge involves ensuring adequate consideration for Internationalization. Presently, the selection of website types and their layout can transcend cultural boundaries, potentially causing surprise or confusion among certain individuals. The effectiveness of humor in transcending geographical and cultural boundaries is limited, and its impact may even be constrained within specific localities. The variations in cues for informality across different groups and cultures. One illustration of this phenomenon is the auditory manifestation that occurs during cognitive processing within the human brain. In the United States, the choice of pause word, whether "uhm" or "ahh," may vary depending on the region. Similarly, in China, the commonly used pause word is "nega." However, when conducting a search and intentionally using an inappropriate pause word for a specific region in order to appear more informal, it can result in an unintended consequence of being perceived as unnatural or strange, thus placing one in the realm of the uncanny valley.

The third subject under consideration pertains to situational awareness. An illustrative case is the distinction between the behavior of an AI system in offline and online settings. The nature of human interaction undergoes modifications when it is disengaged from a network connection. The extent of interaction is contingent upon the cognitive abilities that the user must allocate towards the interaction. If the ability to detect that an individual is engaged in the act of driving is present, it is likely that the responses provided will be of a shorter nature. This topic exhibits a considerable degree of intricacy and subtlety. The AI's ability to detect additional information is advantageous. What is the affective context of the individual's emotional state at that juncture? What is their current emotional state? What types of cues can be discerned from vocal patterns and non-verbal gestures? What insights can be derived from the contextual analysis of a user's journey? What information can be inferred from the user profile?

Let us engage in a discussion regarding errors. The presence of humor exacerbates the issue when the server experiences downtime. Individuals experience a lack of perceived credibility in their interactions. The lack of appropriate behavior exhibited by your AI may result in a loss of trust from users. Instead of utilizing humor, it is recommended to employ empathy as a means of connecting with others. Recognizing and affirming an emotional state is frequently sufficient to engender a sense of comprehension among customers and facilitate the dissipation of negative sentiments arising from a problematic circumstance resulting from an error. After the development and implementation of a personality, how can one ascertain its effectiveness? Let us discuss the assessment of personality traits in relation to achieving success. The objective of this investigation is to determine: Do decision makers exhibit a higher or lower propensity to select products and services? At present, there is a significant prevalence of quantifying social media content through the enumeration of tweets or Instagram images. We would advise against pursuing that course of action of quantifying them poses challenges due to the presence of significant noise. Methods of measurement that we would employ are the application of artificial intelligence in sentiment analysis, assess the brand's strength by employing both qualitative and quantitative survey methodologies and by using AB testing options for conducting comparisons against the baseline. This is an opportune moment to reference the brand's core values. One can obtain word associations pertaining to the altered personality in order to examine its impact on lexical selection, while also maintaining a record of analytical data for uncertain responses. Does the incorporation of personality traits in AI decision-making processes, based on user-provided information, positively influence the level of confidence exhibited by the AI system? Engage in critical reflection and contemplation.

Final words

We as Architects and Interior Designers need to use these AI made design tools to find faster ideas and inspirations rather than relying on it completely for the final design. The final design must be rendered by the human architect for the best user satisfaction.

References

1. Midjourney. https://www.midjourney.com/home/?callbackUrl=%2Fapp%2F
2. Open AI, Dall E2. https://openai.com/dall-e-2/
3. Stable Diffusion. https://stablediffusionweb.com/
4. Dreith, B. 2022. How AI software will change architecture and design. https://www.dezeen.com/2022/11/16/ai-design-architecture-product/
5. Fairs, M. 2022. Artificial intelligence "will empower designers" say Clippings co-founders. https://www.dezeen.com/2021/08/19/artificial-intelligence-empower-designers-clippings-co-founders/
6. Ravenscroft, T. 2022. AI creates "repulsive and strangely compelling" Serpentine Pavilion. https://www.dezeen.com/2022/06/10/ai-designed-serpentine-pavilion/
7. Wiles, W. 2022. Architects can rest easy that AI isn't coming for their jobs just yet. https://www.dezeen.com/2022/11/16/architects-ai-dall-e-midjourney-opinion/
8. Leach, N. 2023. AI is putting our jobs as architects unquestionably at risk. https://www.dezeen.com/2023/02/13/ai-architecture-jobs-risk-neil-leach-opinion/
9. Hebron, P. 2017. Rethinking design tools in the age of machine learning, rethinking design tools in the age of machine learning. Artists + Machine Intelligence | Medium
10. Cassidy, B. 2022. The twisted life of Clippy. https://www.seattlemet.com/news-and-city-life/2022/08/origin-story-of-clippy-the-microsoft-office-assistant

Chapter 5
AI in Visual Communication

AI Taking Down Graphic Designers. Scary?

5.1 Graphic Designer in the Age of AI

Companies throughout the business and design sectors are increasingly innovating with the help of artificial intelligence design tool technologies as AI continues to grow in popularity [1]. As a result, the community's estimation of a graphic designer has shifted because of AI design tools. This chapter is significant because it examines the potential effects of AI-powered design tools on the future of the graphic design profession. As a result, a well-researched study is warranted to describe the problem and examine the potential threat it poses to the public's view of the value graphic designers bring to society. This study will contribute to the field by helping communities and designers understand how AI design tools affect graphic artists' sense of worth to society. This, in turn, will assist communities and designers decide whether to adopt AI design tools.

Graphic design is one field that frequently employs artificial intelligence. Before the advent of personal computers in the early 1990s, most graphic designers created their designs by hand [2]. Because the computerization of the visual design process is an inexorable consequence of the exponential progress in computing power, AI design tools will soon be commonplace in the field [3]. Designers have faith in design to radically alter people's lives [4]. The significance of this design, however, is lost on those who are not designers. Understanding the discrepancy between how designers and the public evaluate a design is essential to the development of the field. Most designers in the field feel that using AI design tools is disrespectful because it diminishes the importance of design and the designers' role in society [4]. To determine whether graphic artists can be influenced by artificial intelligence design tools, it is important to study this area.

Non-designers may not recognize designers' immense worth since designers view design as a transformative instrument to better the human condition. For the future of graphic design, it is worrying that there is a disconnect between how designers and the public regard their work. Therefore, designers feel that AI design tools devalue design and the value that designers bring to society. Thus, we will examine the impact that AI-powered design tools have had on the field of graphic design and how they are being used today. We will uncover description of AI, discuss AI design tools, and examine the perspectives of designers on the topic of AI and graphic design.

AI and graphic design

The phrase "artificial intelligence" must be defined before any tools for designing AI can be established. The concept of teaching computers human abilities and language was proposed by Alan Turing [2]. An example of artificial intelligence, as defined by Karaata [3], is the incorporation of human skills into tools and machines. These tools and machines would then be utilized to perform tasks that humans can do. One branch of computer science, artificial intelligence (AI) aims to mimic human intellect in machines and to automate human tasks using computational means [5]. Artificial intelligence design tools will find a position in the graphic design industry as the process of graphic design continues with computers as technology considerably advances [3]. One way that AI is being applied in business is through AI design tools. When discussing the field of graphic design, the term "artificial intelligence design tools" refers to programs or websites that generate design outputs like logos, page layouts, and website designs using artificial intelligence codes without the participation of a human graphic designer [3]. According to Skaggs [6], AI design tools employ a predetermined algorithm that offers a selection of design specifications like style sheets and compositions, allowing even those who aren't trained as designers to achieve the desired design outcome quickly and easily by adjusting physical parameters like color, font, and size. These are the effects of technology developments that are dramatically impacting graphic designers and the way they execute their jobs [7]. There is a risk that the value of graphic designers will decline as more people gain access to artificial intelligence design tools [2]. Chathurika [1] counters that AI design tools already include helpful algorithms that can supplement the efforts of human designers. Designers might train AI design tools to produce designs in a manner like that of a human designer; this would aid human designers in producing professional, visually stunning final products, but the designers themselves would remain in charge of the project [1]. However, the prospect of machines being able to produce sophisticated design solutions independently of human input in a relatively short amount of time may be cause for concern for the careers of creative designers in the future. According to these justifications, AI has the potential to be a tool that mimics human skills and can replace human designers in a variety of fields, including graphic design.

As a kind of visual communication, graphic design has a place in both academic and professional settings as a distinct design discipline. The process of graphic design moved from being mostly a hand-crafted activity in the early 1980s to a computer-based activity in the 1990s [2]. It's becoming increasingly challenging for designers to adapt to a changing circumstance that affects their societal worth as the process of graphic design continues computers as technology advances dramatically, resulting in influences from new technologies like artificial intelligence design tools [3]. Dorst [8] believes that throughout its evolution from craft to technical professional practice and intellectual discipline, design has always had to find novel means of addressing the increasingly complex problems it has been called upon to solve.

Design, as a natural link between technology and humanity, is ideally placed to contribute to social challenges, according to Dorst [8], who argues that while some complex issues are viewed from either a technological or technocratic perspective, much of the complexity of contemporary problems derives from the human realm.

Designers vs non-designers

Designers consider design to be a "transformative tool" that may be used to better the lives of people [4]. The significance of this design, however, is lost on those who are not designers. For design to progress, it is essential that professionals bridge the gap between what they consider a design's value to be and what the public places on that value [4]. The gap is one of the unavoidable outcomes of AI, as described by Skaggs [6]. Artificial intelligence design tools are seen as disrespectful by the design community because they devalue the profession of design, and the contributions designers make to society [4]. Skaggs [6] argues that the use of AI-powered design tools has lowered the esteem in which graphic designers are held by the public. This is because designers who fall in the middle of the craft-based/analyst field divide tend to be either out of work or earning barely above the poverty line.

Vinh [4], on the other hand, views the potential impact of AI design tools and how they might make design accessible to non-designers from a fresh and optimistic angle. Designers do not like of notion that more people, especially non-designers, will be able to create a piece of design without any assistance from genuine designers [4]. Sites like these offer a wide selection of high-quality, low-cost design options. That's looked down upon and undervalued by designers [4]. While this could be helpful to designers and open doors for others without a design background, not everyone who takes advantage of these changes will go on to become a world-class designer [4]. In conclusion, designers fear that the rise of AI-powered design tools may diminish the respect afforded graphic artists in the eyes of the public.

In conclusion, designers play a pivotal role as translators between people and technology, and design is often regarded as a powerful instrument for enhancing people's lives [4]. One of the most cutting-edge technologies to be used to the design industry is artificial intelligence design software, which is often viewed as a mimic of human designers' abilities to produce unique design outcomes quickly and independently [3]. Furthermore, it offers a plethora of options, making it difficult for human designers to compete. Designers like Swanson [2] worry that the perceived societal worth of graphic designers may decline as more people have access to artificial intelligence design tools previously reserved for designers [2]. However, surprisingly, most non-designers that use AI design tools won't end up being professional, world-class designers [4].

On the other side, these kinds of technologies may end up being helpful in assisting designers in enhancing their productivity and effectiveness. The future of the design business depends on designers closing the gap between how they value their work and how society appreciates their work [4]. Overall, the value placed on graphic designers by the public may shift significantly due to the rise of AI-powered design tools, but this will not be enough to completely threaten the profession of design or lead to its eventual extinction.

Current state

The initial goal of creating an AGI, a machine with human-like capabilities, has not been realized. Thanks to advancements in both data storage and processing power, artificial intelligence has evolved into a simulation of many minds working together to complete a single task, a phenomenon that has been dubbed super-cognition [9].

Big data

The concept of "big data" lacks a concrete definition. Its primary application is with the kind of data that can only be processed by a supercomputer. The definition of what constitutes a "supercomputer"; however, it changes annually. Instead, their scale is a result of the interconnectedness and complexity of their data, connections, and relationships. The ability to collect, store, and analyze massive, linked datasets has resulted in a paradigm shift in the way scientists conduct their investigations. It offers a method that has the potential to transform what it means to learn [10]. Since its inception as a subject of study, artificial intelligence has been approached by creating extensive quantities of code to act as its governing rules. As a form of symbolic logic, the top-down approach is also known as artificial symbolic intelligence. To translate across languages, a computer software needs to be taught the grammatical rules and vocabulary of each language it is expected to handle [11].

Oxford professor of mathematics Marcus Du Sautoy writes in his book "The Creativity Code" how "the deluge of data is the main spark for the new age of machine learning" and how this has altered the trajectory of AI development [12].

Machine learning

Principles of machine learning, which take their cue from the human brain's design, have been around since the 1950s. The perceptron is an artificial neuron that receives data from multiple sources, evaluates it based on a set of weights, and only sends out a signal to another perceptron if the result is greater than a predetermined threshold [12]. A neural network consists of many layers of connected perceptron. The network corrects its data-processing errors by adjusting the weights of the connections among its synthetic neurons [13]. The more layers a neural network has, the more subtle patterns it can pick up on. Deep neural networks are the term for very elaborate neural networks [11]. An underlying meta-algorithm directs machine learning, constructing successive questions and correcting the thresholds of individual artificial neurons and the weights of connections between them based on the data it receives. Because of the abundance of available data from big data projects, neural networks were able to mature into useful tools [12]. It is possible to be misled by the assumption that data is objective and indicative of the truth. One must think about the data's source reliability, and there are always potential biases and blind spots in any analysis [10]. Evaluating the efficacy of a machine-learned system is difficult since the human brain is not wired to assign probabilities.

Belief is the primary foundation of black-box AI, which does not reveal its internal workings to the human operator. It takes a lot of practice with different tests before you can reliably assign probabilities. Artificial intelligence excels because it can interact with data at a far higher rate than humans can [12]. Using human-created labels as training data, AI learns to predict the likelihood that a piece of material satisfies a given criteria and evaluates it against the label it has been given. The AI gradually learns via trial and error to correctly forecast labels for stuff it has never seen before. Supervised learning describes this type of instruction. It is highly dependent on the sophistication of the human-provided, labeled training data [11]. However, these AI-provided datasets may include flaws that teach the AI to attribute causation where none exists. As an illustration, the US military has requested an artificial intelligence that can identify tanks in photographs. The development team collected many images of a tank in different settings and perspectives, as well as images used as references that did not include the tank. The overcast weather obscured the tank images, but the clear weather of the reference photos was missed. The resulting AI can tell the weather apart from the presence of a tank, but not vice versa [12]. As a result of the potential for biases and blind spots in the datasets used to train the AI, a new method had to be created.

Deep learning

When dealing with more generalized problems that lack clear boundaries, unsupervised learning can be used to bypass the need for human-labeled data and the associated risks. The AI can learn to recognize patterns and their associations on an infinite number of things given a large enough dataset, sufficient processing capacity, and sufficient time. Clustering, in which input data is organized into groups with similar characteristics, is an example of unsupervised learning. Reinforcement learning is another method; in this case, the AI is taught by receiving input from its surroundings [14]. Unlike typical machine learning, deep learning can make inferences from unlabeled data without the assistance of a human programmer [15]. To perform complex analyses of incoming data, a neural network must have a multilayer design. A multi-stage process begins with the discovery of basic concepts and then moves on to the identification of the causes of variance that can lead to a conclusion about the content. using them as building blocks to create something more involved. By describing the world as a layered hierarchy of concepts, where each concept is defined in relation to simpler concepts and where more abstract representations are computed in terms of less abstract ones, deep learning can gain immense power and flexibility [16]. Datasets used for training can be provided by other AI systems. Several AIs can work together as either a data provider or a quality checker. Complex networks of different kinds of artificial intelligences can work in tandem [13].

Computer vision

Graphic design is mostly a visual discipline, so we'll talk about how computers interpret images and computer vision separately. Computer vision, as defined by Simon Prince, a professor at University College London who specializes in the study of machine vision and image processing, is the "capacity to extract valuable information from images [17]. That hundreds of different items can be present in any given real-world environment, but that rarely more than a few of them will be in a "typical" position when they are all in full view is something he acknowledges as adding complexity to visual data. It's difficult for computers to do anything as simple as spot object boundaries [17].

Research in computer vision focuses mostly on developing methods to mimic human performance. When it comes to deep learning for computer vision, "most deep learning is used for object recognition or detection," whether that means reporting the presence or absence of an object in an image, annotating the image with bounding boxes around each object, transcribing a sequence of symbols from an image, or labeling each pixel in an image with the identity of the object it belongs to [16]. It has been shown that these modest improvements in artificial intelligence have already had an influence in areas such as "digital photography, visual effects, medical imaging, safety and surveillance, and web search," despite falling well short of the lofty goals of the pioneers of the field [18].

When it comes to the average user's ability to decipher visual content, photography has made the most progress. The public now has access to automated photo editing, image categorization, and even the creation of completely synthetic images. Within the next decade, software will be able to synthesis a snapshot from a list of descriptive terms, eliminating the need for cameras. Alex Savsunenko, CEO of Skylum software, an organization that develops AI-driven photo editing software, was recently interviewed [19]. Google Photos was the first major tool to publicly release that made use of computer vision. With the help of geolocation and picture recognition, this enabled users to search through and organize their photo collections automatically quickly and easily.

There is still a long way to go until we can synthesis images that look as though they were taken in real life from the information gleaned from observing the world around us. To further advance, researchers turned to Generative Adversarial Networks (GANs), which employ two adversarial neural networks to take advantage of deep learning's benefits. While one neural network is being taught to identify, say, cats in real-world photographs, another is attempting to generate phony cat images to test against the trained one [20]. In March 2019, NVIDIA launched a new tool that can make even the most basic sketches look like photorealistic landscapes. "It's like one of those coloring book pictures where you can see the tree, the sun, and the sky all in one image. According to Bryan Catanzaro, NVIDIA's vice-president of applied deep learning research, "the neural network is able to fill in all of the detail and texture, as well as the reflections, shadows, and colors, based on what it has learnt from real photographs [21].

Generating illustrations from a written description

The research company OpenAI has created a computer that can convert straightforward language instructions into high-quality graphics. DALL-E 2 is an AI-based tool that can generate photorealistic artworks from a human-written, free-form textual description. Actions, artistic styles, and different subjects can all be woven into elaborate descriptions. For instance, "an astronaut sunbathing in a tropical resort in space in a vaporwave style" and "teddy bears working on new AI research underwater with 1990s technology" are both examples from the OpenAI blog [22].

In January of 2021, OpenAI released DALL-E, a tool that served as the foundation for the subsequent iteration, DALL-E 2. Thanks to improved image quality, enhanced text understanding, accelerated processing, and the addition of additional features, the latest version can produce even more impressive results. DALL-E is a form of neural network, a computing system loosely based on the interconnected neurons of a real brain. It was inspired by the Pixar robot WALL-E and the artist Salvador Dal. The neural network has been taught to recognize patterns in both visual and written data [22].

OpenAI claimed that their AI could "learn from relationships between objects" in addition to "understanding particular objects like koala bears and motorcycles" thanks to deep learning. Furthermore, "DALL-E knows how to generate that" (or anything else having a relationship to another object or activity) "when you ask DALL-E for a picture of a koala bear riding a motorcycle." Every text prompt in DALL-E 2 can be paired with several different images. DALL-E 2 also allows for the same natural language descriptions to be used for editing and retouching of previously captured images. This tool, dubbed "in-painting" by OpenAI, functions similarly to Photoshop's content-aware fill, allowing users to accurately add or remove objects from a selected area of an image while taking into consideration shadows, reflections, and textures [22].

Examples on the OpenAI blog include inserting a sofa into various locations in an otherwise empty room shot. OpenAI claims that the DALL-E project facilitates visual expression and informs academics about the way sophisticated AI systems perceive and make sense of the world. To create AI that is both beneficial and secure, OpenAI emphasized the importance of this step. Originally founded as a non-profit by high-profile technology figures including Elon Musk, OpenAI is dedicated to developing AI for long-term positive human impact and curbing its potential dangers. To that end, DALL-E 2 is not currently being made available to the public. OpenAI identifies the application could be dangerous if it were used to create deceptive content, similar to current deepfakes or otherwise harmful imagery [22].

It also acknowledges that artificial intelligence (AI) can unintentionally reinforce societal prejudices because of the biases it inherits from its training. While OpenAI improves its safeguards, DALL-E is only made available to a small group of users for evaluation purposes. Users are already restricted from creating any "not G-rated," political, or violent content due to an existing content guideline. Filters and automated and human monitoring systems ensure this remains the case.

The capacity of DALL-E to generate such images initially would be constrained. The training data underwent a thorough removal of any explicit or violent content, resulting in minimal exposure of the model to these concepts. OpenAI was established in late 2015 by Elon Musk, Sam Altman from Y Combinator, and other individuals who provided financial support. However, it is worth noting that Elon Musk has subsequently stepped down from his position on the board of OpenAI. In 2019, the organization underwent a transition to become a for-profit entity, ostensibly with the aim of securing additional financial resources. It is worth noting that the parent company of this organization continues to operate as a non-profit entity. Another project undertaken by OpenAI is Dactyl, which focused on the training of a robotic hand to adeptly manipulate objects through the acquisition of human-like movements that it autonomously learned [22].

5.2 Human-Centered ML

The field of study known as machine learning (ML) focuses on teaching machines how to learn without being explicitly instructed. It's now powering everything from Netflix suggestions to autonomous cars, and it's a potent tool for crafting personalized and dynamic experiences. Even though ML is being used in an increasing number of experiences, UX designers have a long way to go before users truly feel like they have mastery over the technology. Like the mobile revolution and the web before it, ML will force us to reevaluate, reorganize, relocate, and ponder new options for every experience we create. Our goal with the "Human-Centered Machine Learning" (HCML) initiative in the Google UX group is to direct and concentrate that discussion. Using this lens, we examine a wide range of goods to see how ML may address human problems in ways that are both practical and novel. To ensure that ML and AI are developed in accessible ways, our team at Google works with UXers across the firm to help them become familiar with fundamental ML concepts, learn how to incorporate ML into the UX toolkit, and more.

If you are just getting started in the field of ML, you might be feeling daunted by the vastness of the field and the number of possible directions for innovation. Do not rush things and give yourself some time to adjust. Adding value to your team doesn't need you to start from scratch. For designers who are just starting out in the world of ML-driven products, we've outlined a seven-point road map. These guidelines, derived from our experience working with Google's UX and AI teams (along with a good dose of trial and error), will help you prioritize the user experience, test hypotheses rapidly, and capitalize on the opportunities presented by machine learning.

Do not assume machine learning will determine which issues need fixing

There is a lot of buzz about machine learning and AI now. Product strategies that treat ML as an end, rather than the starting point for solving a real problem, are becoming increasingly popular among businesses and product teams.

That is ok for testing the waters or discovering the limits of a technology; in fact, it is a great way to generate ideas for new products. Nonetheless, if you are not focused on satisfying a human want, you'll end up creating a complex solution to a nonexistent or extremely limited problem. Therefore, our first argument is that you still must put in the same amount of effort as before to identify human requirements. Ethnography, contextual inquiries, interviews, deep hanging out, surveys, reading customer support tickets, logs analysis, and becoming close to people are all part of this process. ML won't figure out what issues need fixing. To be precise, we have yet to define it. UX professionals have the resources necessary to lead their teams in any technological environment.

Determine if there is a novel approach to the issue that ML can take

Once you know what problems you wish to solve, you can evaluate whether ML offers any special solutions. It's not hard to find real issues that can be solved without resorting to ML. Determining which experiences need ML, which are significantly improved by ML, and which do not benefit from ML or are even worsened by it is currently a challenge in product development. You don't need ML to make things feel "smart" or "personal." Don't be misled by thinking that ML is required for those.

To aid teams in grasping the benefits of ML for their specific use cases, we've developed a series of activities. These activities accomplish this by exploring in depth the mental models and expectations people might bring into interaction with an ML system, as well as the data requirements for such a system. We utilize the following three examples of exercises to get teams thinking about the use cases they are addressing with ML. Give an account of how a hypothetical human "expert" might complete the assignment as it is right now. What kind of feedback would you give to your human expert if they performed this work so that they could get better at it next time? Follow this procedure throughout the confusion matrix's four stages.

What are the user's expectations if a person were to carry this out?

People's preconceived notions about an ML-powered product can be uncovered by devoting just a few minutes to answering these questions. They work just as well as user research stimuli as they do as discussion starters for product teams. When we get to the part about establishing labels and training models, we'll talk a little bit more about these, too. We next use a convenient 2×2 grid to map out all the team's product ideas after these activities and some more sketching and storyboarding of specific products and features.

This helps us identify the most promising ideas, discard the less promising ones, and determine the extent to which certain concepts rely on machine learning. If you aren't already including engineering as a partner in these discussions, now is a fantastic moment to do so, so that they can provide input on the ML realities of these ideas. In the top right corner of the above matrix, you'll want to prioritize whatever has the highest user impact and is uniquely enabled by ML.

Fake it with your own experiences and magic tricks

The process of prototyping poses a significant challenge in the context of machine learning systems. It is not feasible to expeditiously develop a prototype that possesses a semblance of authenticity if the primary value proposition of the product lies in its utilization of the user's personal data to customize the experience. Furthermore, delaying the implementation of substantial architectural modifications until the completion of a fully designed machine learning system may result in a missed opportunity. Nevertheless, the incorporation of participant examples and Wizard of

Oz studies are two user research methodologies that can prove to be advantageous. The inclusion of participants' personal data, such as personal images, contact lists, and recommendations for music and movies, in user research utilizing prototypes, offers several advantages. It is imperative to ensure that comprehensive information regarding the utilization and disposal of test subjects' data is provided to them. This can also function as a casual task for the group to accomplish prior to the commencement of the session (as individuals tend to enjoy discussing their preferred films).

In this way, you can model both correct and incorrect system behavior. To test how the user reacts and what she assumes the system's error was, you may, for instance, mimic the system recommending the wrong movie. Rather than relying on hypothetical cases or abstract explanations, you can more accurately weigh the costs and benefits of different alternatives with this method. The second method, known as Wizard of Oz research, is also effective for testing unfinished ML products. Wizard of Oz studies were once widely used, but their popularity as a user research technique has waned in the past 20 years. So, they've returned.

For those who may have forgotten, in Wizard of Oz research, participants engage with what they think is an autonomous system but is directed by a human (usually a teammate). Interacting with a "intelligent" system can be "simulated" by having a colleague act out the behaviors of an ML system, such as chat responses, calling suggestion, or movie recommendation. These interactions are crucial for directing the design because users are more likely to create an accurate mental model of the system and adapt their behavior accordingly when given the opportunity to interact earnestly with what they perceive to be an AI. The system can learn a great deal from how people use it and how they adjust and second-order interactions with it.

Compare the benefits and drawbacks of false positives and false negatives

There will be inaccuracies in your ML system. It's crucial to recognize the signs of these flaws and think through how they can influence the product's reception. The confusion matrix was referenced in one of the questions above. This fundamental idea in ML characterizes the outcomes of both successful and unsuccessful ML implementations.

Machine learning (ML) systems exhibit a uniform approach in handling errors, whereas individuals possess varying criteria and expectations. In the context of a classifier designed to distinguish between individuals and trolls, misclassifying a human as a troll would constitute a system error. The system renders assessments without taking into account the emotional state or personal history of the user, and it formulates sweeping generalizations without adequately considering the cultural ramifications. It doesn't consider that users can be more angered by being mistakenly labeled a troll than by trolls being mistakenly classified as individuals. That said, it could just be that we have a bias for humans in general. For ML to work, you must consciously balance the system's accuracy and recall. In other words, you'll have to weigh how crucial it is to maximize the number of correct responses while accepting a larger number of incorrect ones (optimizing for recall) versus how crucial it is to minimize the number of incorrect responses while accepting fewer correct ones (optimizing for precision). If you type "playground" into Google Photos, you can get results like this.

Plan for collaborative study and change

The best ML systems are always adapting to new data and users' evolving mental models. People's actions within these systems shape the future results that others will perceive. The models will be updated in response to the new behavior of the system's users, and so on and so forth. Conspiracy theories emerge when people construct flawed mental models of a system and then run into issues trying to influence the outputs to conform to those assumptions. Users should be directed with well-defined mental models that inspire them to provide constructive criticism.

Even though ML systems are educated using preexisting data, they will respond to novel inputs in ways that are not always predictable in advance. Since this is the case, we must modify our methods of user feedback and survey administration. This requires coordinated longitudinal, high-touch, and broad-reach research to be planned early in the product lifecycle. As the number of users and use cases expands, you'll need to set aside more time to sit with people as they interact with these technologies and observe how their mental models change in response to each success and failure. As UX professionals, we also need to consider how we can collect real-time user feedback across the product's whole development cycle to train and optimize the ML algorithms. Good ML systems can begin to be differentiated from outstanding ones by designing interaction patterns that enable providing feedback simply and immediately demonstrating its benefits.

Use proper labels when training your algorithm

Wireframes, mockups, prototypes, and redlines are some of the most common deliverables among UX professionals. Surprise: there is a limit to how much we can express in terms of a user experience that is augmented by machine learning. This is where the concept of "labels" becomes useful. Machine learning relies heavily on labels. Individuals are employed to examine vast quantities of content and assign tags, such as "is there a cat in this photo?" After enough images have been manually classified as "cat" or "not cat," you will have a dataset from which to educate a model to recognize cats. More precisely, the ability to determine whether a cat is present in an unseen image with a certain degree of certainty. Not rocket science, is it?

The difficulty arises when trying to apply a model to anticipate user reactions to content, such as whether they will find a given article interesting or a given suggested email reply significant. However, training models takes a long time, and it can be prohibitively expensive to acquire a dataset correctly labeled; also, incorrect labeling might have a major effect on the success of your product. Follow these steps to proceed: To get started, it's important to make informed assumptions and have those assumptions discussed with a wide range of partners. These should be phrased in the manner "we assume _____ consumers will choose _____ and not _____ __ in _____ scenarios." The next step is to swiftly implement these hypotheses into a prototype to begin collecting user data and iterating. To train your machine learner effectively, you'll need to track down subject-matter experts who can speak authoritatively on the topics related to your forecasts. Our advice is to either bring on several new people to fill this position or to train an existing employee to do so. On our team, they are referred to as "Content Specialists."

You will be able to tell at this stage whether your presumptions feel more grounded. A second stage of validation using examples picked from real user data by Content Specialists is essential before spending heavily in large-scale data collecting and labeling. The prototype being tested by your users should be of good enough quality that they believe they are dealing with a real artificial intelligence. Once you have proof that your AI is functioning as intended, have your Content Specialists build a library of examples that are good representations of the output you want to see from the system. These examples provide a road map for data collecting, a robust set of labels to kick off model training, and a framework for creating large-scale labeling processes.

ML is a creative process; broaden your UX team to incorporate its members

A professional specializing in user experience (UX) encountered suboptimal "feedback" due to micromanagement. The presence of the critic appears to be in close proximity, positioned to identify and highlight any errors that may be committed. It is important to maintain a mental image of the aforementioned scenario and ensure that one does not project such a demeanor towards the engineers. There exist numerous potential methodologies for addressing machine learning challenges. Consequently, as a user experience professional, it is advisable to exercise caution in providing overly specific instructions at an early stage, as this may inadvertently restrict the creative input of engineering colleagues. Rely on individuals to employ their intuitive faculties and motivate them to engage in experimentation, even if they may exhibit reluctance to conduct user testing prior to the establishment of a comprehensive evaluation framework. Machine learning represents an engineering process that is characterized by heightened levels of creativity and expressiveness, surpassing the conventional norms to which we are typically accustomed. The process of training a model is often characterized by its slow pace, and the available visualization tools are still lacking in quality. Consequently, engineers are frequently required to rely on their imaginative abilities when fine-tuning an algorithm. In fact, there exists a methodology known as "active learning" wherein engineers manually adjust the model's parameters after each iteration. The primary responsibility assigned to you is to assist individuals in making optimal user-centered decisions throughout the entire process.

Motivate individuals by presenting illustrative instances of an exceptional encounter, encompassing visual aids such as presentations, anecdotal accounts, visionary videos, prototypes, and excerpts from user research. Facilitate their proficiency in comprehending the objectives and discoveries of user research, and gradually introduce them to the realm of UX critiques, workshops, and design sprints, thereby fostering a more profound comprehension of your product's guiding principles and experiential objectives. The longer the duration allocated for their acclimation to the iterative process, the greater the enhancement of both the machine learning pipeline and their ability to impact the product.

5.3   Danger to UX by AI

Is there a possibility that AI will render UX design irrelevant in the future?

"The optimal approach is consistently one of augmentation rather than substitution"

The commonly proposed scenario revolves around the replacement of repetitive tasks by artificial intelligence (AI) in the initial stages. Subsequently, Generative Adversarial Networks (GANs) commenced producing a wide array of outputs. In light of the computer's ability to generate a thousand designs within a single second, the necessity of employing a designer may be called into question. Right? Let us commence with an illustrative instance. Deep Blue, the chess application that initially garnered widespread attention for its victory over the world's top chess player, is widely recognized. Since its inception, artificial intelligence (AI) has achieved remarkable milestones, such as surpassing the skills of the top Go player and demonstrating superior performance in competitive video games against human opponents. However, are you aware of individuals or entities that have successfully outperformed artificial intelligence systems? The integration of human and artificial intelligence (AI) systems. Currently, the highest-performing chess systems, capable of surpassing any existing AI, consist of a combination of human and artificial intelligence components. Both the human brain and an artificial intelligence (AI) system exhibit the tendency to employ cognitive shortcuts. Various approaches are employed to carry out these tasks. They are more effective at compensating for each other's areas of weakness. The optimal approach involves augmenting existing systems rather than completely replacing them. According to IBM CEO Ginni Rometty, she expresses a preference for the term "augmented intelligence" over the acronym "AI."

Although it is unlikely that AI will completely replace or render UX obsolete, it is anticipated that significant changes will occur in the field. In the aforementioned scenario, it was discussed that Generative Adversarial Networks (GANs) have the capability to generate a substantial number of designs, specifically 1000 designs per second. This phenomenon can indeed occur. However, the process of automated generation does not necessarily guarantee quality. In previous years, a company named The Grid emerged with the proposition of eliminating the necessity for website design. The outcomes it produced were disappointing. However, one might inquire whether the situation could potentially improve if an alternative company were considered. Google attempted a comparable endeavor. It is possible that you are familiar with the practice of conducting a comparative analysis of 42 different shades of blue in order to identify the optimal hue that elicits the highest response rate. The endeavor yielded positive results. However, their attempts to extend the application of analytics-based design beyond rudimentary elements were consistently met with obstacles. Therefore, what is the underlying reason for this phenomenon? In order to enhance the performance of artificial intelligence, it is imperative to improve the user experience. A reciprocal advantage exists between both parties, which operates in a cyclical manner. As the utilization of artificial intelligence (AI) becomes more prevalent, the machine learning (ML) model generates an increasing amount of valuable data. The utilization of the aforementioned data to train the new model results in an enhanced functionality of artificial intelligence. Machine learning models are increasingly emerging and providing unnecessary advice and tasks, thereby exacerbating confusion rather than resolving problems. The importance of improving user experience (UX) is increasingly recognized. The creation and refinement of a more optimal user experience (UX) is undertaken. The utilization of artificial intelligence is increasing, thereby initiating a recurring cycle.

If the cyclical nature of UX indicates an ongoing demand for its services, what transformations can be anticipated in the nature of the UX profession? Similar to many other occupations, the elimination of tedious, repetitive, and monotonous tasks is occurring. The process of design will undergo automation. In the aforementioned scenario, the focus was on the remarkable capability of Generative Adversarial Network (GAN) machine learning models to generate a staggering number of 1000 designs per second. The existence of the mentioned entity has already been established. The role of the new UX designer will shift towards that of a curator rather than a creator of designs. The advancement of design tools and the maturation of user experience (UX) design have already initiated this process. There exists no justification for continuously undertaking the task of redesigning identical widgets throughout the entirety of one's professional trajectory. The existence of systems designers serves as substantiating proof of this claim. The design of a component is conducted once, after which it is utilized and tailored by all users within the system. The aforementioned statement should remain valid. Let us delve into a comprehensive discussion of the three primary domains associated with design, namely design, research, and management.

Design

The focus of your designs lies primarily on empathy, without introducing any novel concepts or ideas. One of the challenges associated with machine learning models pertains to comprehending the concept of humanness. How can one determine the means by which an application can respond to the present context and emotional state of the user? One area of current advanced research is focused on ensemble models. The fundamental concept posits that machine learning models exhibit proficiency in specific tasks. Consequently, amalgamating multiple models and incorporating an additional model to determine the optimal model for a given task enhances the overall resilience of the system. The design of the aforementioned item is of utmost importance. Each alteration in context necessitates the utilization of a distinct machine learning model. In order to establish a comprehensive understanding of the situation, I previously discussed the importance of discerning appropriate moments for the delivery of humorous remarks. The answer to that question is contingent upon the specific context in which it is being considered. The significance of context is also evident in situations where the model recognizes its errors. In such cases, it becomes necessary to devise a mechanism for the model to acknowledge and accept its mistakes.

The user interface of a device is influenced by the context in which it is used. It is essential to ascertain the specific device being utilized by the individual and discern the distinctions between said devices. There is a software tool known as Applitools (https://applitools.com) that facilitates cross-platform testing. How does the contextual interpretation vary depending on the device being utilized? It is imperative to stay updated on the latest device releases due to the evolving nature of user interface (UI) modifications. Advancements in technology have resulted in the introduction of novel devices, which in turn offer a range of additional functionalities. Familiarize oneself with the available features and tailor the experience accordingly. An example of a voice-activated virtual assistant is Amazon Alexa. Upon its initial release, the device functioned solely as an audio output device. The device now possesses a visual display and the capability to engage with various screens within the household. The design of the interaction is necessary due to the historical effectiveness of AI across various domains. The determination of this outcome is contingent upon the analysis of historical data. The predictive capacity of the system is evident, although its ability to acclimate to a novel paradigm, such as the introduction of a new product, is characterized by a protracted adjustment period. It is imperative to strategically plan and develop these experiences.

It should be noted that artificial intelligence (AI) exhibits limitations in effectively executing transitions. Due to the imperative of concentrating models on a highly specific domain, it becomes necessary to transition from one model to another in order to comprehensively address the entirety of the user journey. The success of this transition will depend on your actions and decisions. Furthermore, in the event of a change in modality, appropriate design adjustments will also be required. For instance, when a user switches from a laptop to a mobile device, it becomes challenging for a single model to seamlessly transfer all the necessary information to the other model. Consequently, the determination of what aspects are crucial for facilitating this transition must be made as an integral part of the design process. Artificial intelligence systems often struggle to handle edge cases effectively. As previously discussed, accessibility can be regarded as a collection of exceptional scenarios that can be effectively addressed through the implementation of a machine learning model. Failure to prioritize accessibility can expose organizations to legal action and result in the loss of approximately 12–15% of their customer base. Furthermore, it is important to consider that disregarding these disabilities may introduce additional interference to your model. This is particularly relevant when your model involves interaction recognition, as muscular disabilities have the potential to disrupt the accuracy of such models. Similarly, cognitive disabilities can adversely affect the reliability of data answers. Given the aforementioned reasons, it is imperative to further enhance the differentiation of accessibility personas beyond their previous levels.

Research

The initial step for a researcher entails reviewing the collected data intended for model training purposes. Is there a correspondence between the data and the user's intended purpose? What was the rationale behind the collection of data and at what point in time was it collected? Could the impact of this have any implications on the accuracy of the data used for training the model? It is imperative to exercise caution and remain vigilant for any potential gaps or inconsistencies within the dataset when conducting a comparative analysis with the field data. Alternatively, the model may exhibit a high level of accuracy yet fail to effectively cater to the specific customer demographic that the app intends to reach.

For researchers in the field of user experience (UX), it is evident that despite the evolving nature of their work, certain aspects remain consistent. The user journey holds significant importance in the context of user experience. Currently, the situation is more pronounced than it was in the past. Machine learning serves the purpose of automating mundane tasks. The objective of this inquiry is to ascertain the means by which individuals can engage in activities that align with their personal passions. It is advisable to refrain from automating the domains that individuals derive pleasure from. In order to identify those specific areas, conducting research is necessary. As mentioned earlier, machine learning encounters challenges in addressing human aspects, necessitating insights into user motivations and pain points, which are typically provided by the UX researcher. One of the most significant challenges faced by artificial intelligence (AI) pertains to the issue of trust. In order to effectively execute appropriate actions at the appropriate moments, it is imperative to possess a comprehensive understanding of the user journey and account for the various contextual shifts that impact said journey. In order to establish trust, it is imperative for a researcher to identify the critical stages within the user journey. Where is the requisite level of accuracy expected? It is acceptable if the model's accuracy rate is limited to 70 or 80%. The ability to discern the relative significance of various stages within the user journey will enable developers to determine the areas that require their attention and effort.

It is imperative for researchers to possess comprehensive knowledge of various personas. Historically, the prevailing approach has been to condense personas into three or four archetypes. The necessity for modification arises due to the limited scope of the artificial intelligence system. In order to effectively train the model, it is imperative to collect data from populations that exhibit similarities to the target population under consideration. Alternatively, potential issues may arise due to the presence of data gaps that correspond to the inclusion of your customers. In order to achieve internationalization, it is imperative to develop distinct models for each target market. In order to gain a comprehensive understanding, it is imperative to ascertain the dissimilarities between these target markets. In what manner should the developers modify their approach to model construction in order to accommodate the unique characteristics of each group? The responses to these inquiries must be incorporated into your written reports.

Management

Firstly, it is essential to conduct a benchmark analysis of the existing product. Conduct a comparative analysis of rival products. Numerous methods that have been discussed pertain to adapting to changes in processes. Given the inherent opacity of machine learning models, a novel approach has emerged wherein metrics are utilized to compare against a baseline. First, it is crucial to determine the significant metrics before proceeding to conduct a benchmark analysis. Models can quickly become non-functional. Libraries undergo rapid updates, making it potentially challenging to activate an outdated model for comparative purposes. Therefore, it is advisable to undertake this task promptly. In addition, the management of the process entails the inclusion of supplementary aspects that require attention. One of the most significant concerns is likely to be ethics. Is it advisable for you to engage in the construction of it? The discussion does not pertain solely to the feasibility of constructing the problem without the utilization of artificial intelligence, although it is imperative to ascertain this aspect as well. Are there any unintended consequences associated with the construction of the product? Will this product be the optimal choice for the user? In what manner do you exert influence over the actions of the users?

In addition, it is important to acknowledge the emergence of novel risks associated with machine learning models, particularly their susceptibility to deception. The potential for certain users to deceive the model has the potential to result in a suboptimal experience for other users. It is imperative to ensure that the model is not susceptible to exploitation by comprehensively understanding the input data and conducting a thorough comparison with the output data. This enables malicious users to identify exploitable shortcuts within the model. Additional aspects of security encompass the training data, data sources, algorithms that interface with the models, as well as the models themselves. It is advisable to possess the knowledge and skills required to restore a system to a previously established functional state in the event of an error. Additionally, it is crucial to implement measures that prevent the model from modifying its own structure or behavior. The consideration of transparency is important, even in cases where users may not explicitly request it. It is plausible that the matter has not yet been given due consideration, and once it is, the perspective may swiftly shift in the event that the application lacks transparency. The manner in which the process is disclosed will vary depending on the application, as well as the quantity and nature of the information that one intends to provide. It is imperative to incorporate within the design process a consistent consideration regarding the disclosure of data sources and processing methods whenever data is utilized to address inquiries or accomplish objectives. Given that artificial intelligence (AI) is designed to perform mundane tasks on behalf of users, it is beneficial to enhance transparency by explicitly notifying users when assistance has been provided.

The field of AI safety is increasingly being addressed by professionals in the field of user experience (UX). Various forms of safety exist in the context of businesses' utilization of artificial intelligence (AI). It is crucial for businesses to acknowledge the inherent unpredictability of AI systems. Consequently, if an application incorporating AI is employed within government or regulated sectors such as banking, it may potentially give rise to complications with regulatory authorities. In the context of mission-critical systems, it is imperative to acknowledge that relying solely on the testing of 1000 iterations with the expectation that one will yield a successful outcome is not a viable approach. Consequently, designers are compelled to develop safety scaffolding measures to establish operational boundaries for the machine learning (ML) model. Ensuring AI safety for users entails the recognition of contextual cues, particularly in situations where providing an incorrect answer could potentially result in harm. To address this concern, it is imperative to incorporate within the user experience a mechanism that allows for the transfer of control to human intervention or the complete termination of the interaction. There is a growing imperative to acknowledge and address human biases. Given that data is derived from individuals, it is important to acknowledge that individuals possess inherent biases. Thus, the data is. Bias can potentially infiltrate various aspects of data, including training data, labeling data, data collection methods, data cleaning processes, and data output formats. One effective approach for verification involves transforming the identified research needs into narrative form. Examine the data ingestion process at each stage of transformation to ascertain its alignment with the narrative's underlying purpose. The identification of bias is facilitated.

Another aspect that was not discussed is Artificial General Intelligence (AGI). For those who may not be acquainted with the concept, it refers to the portrayal of artificial intelligence (AI) in films as possessing the ability to engage in autonomous cognition, in contrast to the current state of narrow AI, which excels at performing specific tasks. Given the considerable controversy surrounding the feasibility of achieving Artificial General Intelligence (AGI), it is pertinent to emphasize the imperative of prioritizing human-centered design in this context. The incorporation of AI transparency, ethics, and safety into the development of ML models is crucial. Despite the absence of the AGI movie scare, it is evident that machine learning possesses immense power. The process of comprehending the underlying purpose behind data collection and subsequently transforming it into actionable insights necessitates the utilization of more advanced tools than those currently available. Improving the design aspect is instrumental in achieving the desired outcome.

Final Words

We as graphic designers need to leverage the potential of graphic AI design tools for making the process of design faster and at the same time intervening wherever necessary for an augmented final product.

References

1. Chathurika, H. 2019. Future artificial intelligence in design. *UX Planet.* 14 October. https://uxplanet.org/future-of-artificial-intelligence-in-designce8f4e7a2adc.
2. Swanson, G. 2019. Refuturism. *Dialectic: A Scholarly Journal of Thought Leadership, Education and Practice in the Discipline of Visual Communication Design* 2 (2). https://quod.lib.umich.edu/d/dialectic/14932326.0002.202/--refuturism?.
3. Karaata, E. 2018. Usage of artificial intelligence in today's graphic design. *Online Journal of Art and Design* 6 (4): 1–17. https://www.researchgate.net/publication/331431169_Usage_of_Artificial_Intelligence_in_Today%27s_Graphic_Design.
4. Vinh, K. 2019. Welcome to the designers' club-keep out. *AIGA Eye on Design.* 11 December [Online]. Available at: https://eyeondesign.aiga.org/welcome-to-thedesigners-club-keep-out/ [Accessed 27 June 2020].
5. Luger, G.F. (2009). *Artificial Intelligence: Structures and Strategies for Complex Problem Solving.* Boston: Pearson. Cited in: Karaata, E. 2018. Usage of artificial intelligence in today's graphic design. *Online Journal of Art and Design* 6 (4): 1–17. http://www.adjournal.net/articles/64/6410.pdf.

6. Skaggs, S. 2019. Stage 1.X—Meta-design with machine learning is coming, and that's a good thing. *Dialectic: A Scholarly Journal of Thought Leadership, Education and Practice in the Discipline of Visual Communication Design* 2 (2). Available at: https://quod.lib.umich.edu/d/ dialectic/14932326.0002.204?view=text;rgn=main [Accessed 22 June 2020].

7. Amatullo, M. 2019. Designing the future we want. *Dialectic: A Scholarly Journal of Thought Leadership, Education and Practice in the Discipline of Visual Communication Design* 2 (2). https://quod.lib.umich.edu/d/dialectic/14932326.0002.201/--designing-the-fut ure-we-want?rgn=main;view=fulltext;q1=artificial+intelligence.

8. Dorst, K. 2019. Design beyond design. *She Ji: The Journal of Design, Economics, and Innovation.* 4 May [Online]. Available at: https://www.sciencedirect.com/science/article/pii/S24058 72618300790 [Accessed 21 June 2020].

9. Manovich, L. 2018. *AI Aesthetics.* Manovich.Net. Retrieved July 25, 2020, from http://man ovich.net/index.php/projects/ai-aesthetics.

10. Boyd, D., and K. Crawford. 2011. Six provocations for big data. *SSRN Electronic Journal.* https://doi.org/10.2139/ssrn.1926431.

11. Lewis-Kraus, G. 2016. The great A. I. awakening. *The New York Times.* 14 December. Retrieved July 25, 2020, from https://www.nytimes.com/2016/12/14/magazine/the-great-ai-awakening. html.

12. Sautoy, M.D. 2019. *The creativity code: How AI is learning to write, paint and think.*

13. Spacey, J. 2016. *What is a neural network? Simplicable.* Retrieved July 25, 2020, from https:// simplicable.com/new/neural-network.

14. Joshi, A.V. 2020. *Machine learning and artificial intelligence.*

15. NVIDIA. 2020. *Deep learning NVIDIA developer.* Retrieved July 25, 2020, from https://dev eloper.nvidia.com/deep-learning.

16. Goodfellow, I., Y. Bengio, and A. Courville. 2016. *Deep learning.* Retrieved July 25, 2020, from https://www.deeplearningbook.org/.

17. Prince, S.J.D. 2012. *Computer Vision: Models, Learning, and Inference.* Cambridge University Press.

18. Szeliski, R. 2011. *Computer Vision: Algorithms and Applications.* Springer.

19. Lyndersay, M. 2019. *How AI is changing photography Tech News TT.* Retrieved July 25, 2020, from https://technewstt.com/bd1169/.

20. Beckett, J. 2017. What's a generative adversarial network? Google researcher explains. *The Official NVIDIA Blog.* Retrieved July 25, 2020, from https://blogs.nvidia.com/blog/2017/05/ 17/generative-adversarial-networks/

21. Salian, I. 2019. Gaugan turns doodles into stunning, realistic landscapes|nvidia blog. *The Official NVIDIA Blog.* Retrieved July 25, 2020, from https://blogs.nvidia.com/blog/2019/03/18/ gaugan-photorealistic-landscapes-nvidiaresearch/.

22. Aouf, R.S. 2022. *OpenAI's DALL-E 2 generates illustrations from written descriptions.* https:// www.dezeen.com/2022/04/21/openai-dall-e-2-unseen-images-basic-text-technology/.

Chapter 6
AI in Interaction Design

AI Interacting Efficiently with the User. What?

6.1 Interacting with AI

Between 10 and 60% of occupations could be automated soon. It is all happening far more quickly than we anticipated. With the rapid expansion of the machine learning and artificial intelligence sectors, discussions of a universal basic income have picked up speed. Even if robots end up doing all our jobs, we humans must find a way to support ourselves. Hence, let us levy a fee on the machinery and distribute the proceeds to the people. There are significant societal and cultural issues with either strategy, but we should not think that far forward just yet. Humans will no longer have to perform routine, uninteresting tasks since machines will do it for them. Self-service escalators, taxis that drive themselves, and other similar innovations are already here. Sensors and AI will eventually replace all the mundane, repetitive work that humans have traditionally done. As a result of their ability to learn and improve over time, they can easily outperform their human competitors. As a result, fewer clerical errors and vehicular mishaps will occur. Dull labor will be automated away. Bored, repetitive work is what robots are built to do. The use of AI in healthcare is exploding. According to a BBC story released in January, AI is already superior to doctors in the field of breast cancer diagnosis. And these experts are highly specialized and paid accordingly. To "correct" and "train" the AI, humans will always be required [1].

AI assisting doctors

AI can be a helpful addition to medical teams, but it will never replace human doctors (for the most part). Moreover, patients require that "human touch" during their treatment, which can only be provided by these professionals. Thus, robots will never replace human doctors. They will rely on machines far more than they do now. Most of their tasks will be completed automatically by AI. One of the things that is already happening is the scanning and analysis of x-rays, but doctors with robot arms, or even surgical nano-robots, are on the horizon.

M. H. Akhtar and J. Ramkumar, *AI for Designers*,
https://doi.org/10.1007/978-981-99-6897-8_6

AI-driven user experience (UX) research and design

We will test how much of the research and design processes can be automated and how does that affect the UX procedure? Let us evaluate which user experience roles (or steps in the UX process) artificial intelligence might one day be able to fill. It is safe to presume that AI will not be able to conduct one-on-one interviews, lead seminars, or moderate focus groups with humans anytime soon. Some attendees may even find being led by a machine to be disturbing. Thus, it is safe to assume that the researcher's ability to interact with the user directly will be immune to AI (for now). What remains of the procedure, then, deserves our attention.

Initiating research: quantitative data analysis

AI is already quite good at analyzing large datasets for patterns and drawing inferences from them. A move to using AI to conduct quantitative research is possible. Perhaps the most apparent thing to do would be to search for patterns in the generated dataset. Why don't we teach the AI to make up its own polls and questionnaires? Yes, it will take some time for a human to educate the AI to accomplish that well, but many standard research problems are universal. Artificial intelligence (AI) might therefore take over quantitative research in its entirety, initially with minimal input but subsequently unaided and guided. Robots are competent data analysts.

The next step: qualitative studies

Human intuition is especially important in qualitative research. After all, we need user empathy to fully comprehend their actions and issues. Yet, the pool of unique individuals with the necessary skills, with a few notable exceptions, is quite small. If we rate the user's proficiency on a scale from 1 to 100, we can still train AI to provide a suitable response. In addition, we can directly poll people to see if the AI correctly interpreted their intent. Thus, it would be possible for the users (with some moderation) to train the AI to better comprehend them. This is already happening in the realm of social media, where AI is constantly monitoring our preferences and anticipating our next clicks.

Robotics and qualitative research

It should be only somewhat more challenging to employ AI in qualitative research. It will need constant human monitoring for a much longer time than quantitative metrics. Artificial intelligence (AI) can provide both the cards and the findings for card sorting, and if a tablet (touch and drag) interface is used in place of real cards, the entire process can be carried out digitally, without any involvement from the researcher. Teaching a self-driving AI is like click-testing or analyzing click and navigation patterns in prototypes. At first, you have a highly skilled driver take the wheel so that AI may learn from their example. As a result, it will be able to analyze user behavior and identify any discrepancies that require further investigation and customization based on the user's level of expertise. Users' actions can be observed and compared to the desired result to conduct task-based usability studies. The user's internal dialogue, if any, during the test can also be recorded and analyzed for relevance (and pinpointed to the moment on the screen, when the words were said).

Organizing data using cards

AI cannot only analyze the results of card sorting but also generate the cards themselves with some guidance. Evaluation using heuristic criteria is beyond the scope of this article. What happens when people spend a lot of time and effort on a lot of different projects over time. In contrast, excellent practices or established guidelines can be fed into the AI. Again, AI can take care of this with sufficient training. If AI is properly trained, it can easily pass the system usability scale (SUS). While it's essential to identify your audience, doing it without the aid of make-believe companions can be accomplished in a fraction of the time. Using our predetermined criteria or data from prior user-workshops, AI may create names for them, retrieve photographs of them from one of those UI-faces sources, and then automatically prefill the personas. This demonstrates that the entire procedure can be completed in a matter of seconds [1].

Design

How AI may aid in the creative process? Although the term "creative" is often used to describe the design process, originality can only be achieved through the user interface and the micro-interactions. The remaining work entails generating flows and wireframes in accordance with established best practices and existing research, leaving little room for original thought. Ultimately, the goal is efficacy, not novelty alone.

Use flowcharts

Card sorting can be used to automatically build flowcharts. A human should presumably draw the major branches of the flow diagram if we did not know algorithms and flowcharts. In this way, AI may prefill all the branches based on similarities in the data. Login and registration, two seemingly elementary phases of a process, can often be depicted entirely by a computer-generated flowchart.

Wireframes

Intricacies aside, if a flowchart already exists, each node can be developed into a wireframe, albeit at a low fidelity. Those that require little thought or input from a human being can be prefilled and saved for later use. If the flow is more complex, the AI will need to learn many distinct ways to accomplish the same goal. The app is being used to make a purchase for the customer. The AI must be given background on the variety of possible outcomes of the process to grasp its complexity. How it stacks up against the other thousands of e-commerce apps out there? Even though those businesses will not share their data with us, we can still have our bot make purchases at each shop without needing to verify its actions.

AI-generated wireframes with thoughtful user input

Many wireframes (especially the simple ones) can be made automatically. If we feed the system enough preexisting flows (in the form of wireframes), it will propose a checkout process (the whole process) for us, leaving us to do nothing more than tweak it to our liking.

High-fidelity design

In this case, there are two options for the user interface design. It is still ideal to allow a human to come up with the truly innovative, product-specific style that customers will remember the product by (with a good eye do it). Although AI can mimic human-made UI, it will not be original. Moreover, if we are looking for something delightful, we may instruct the AI to research existing delightful designs on Dribbble [2] and create something similar. Yet without human input, all projects would start to look the same since the machine would not dare to be creative. It would make use of whatever is currently trending, like the numerous AI-designed Neumorphism [3] examples on Dribbble. AI can still be useful in this scenario by detecting grid-misalignments, counting the number of fonts on the artboards and throughout the document, etc. One of the most troublesome aspects of any user interface is the alignment of text on buttons, which may be checked optically with this tool.

Develop superior button designs

AI can design better user experience buttons based on the most liked buttons having all best features in terms of form, color, placement, etc. AI will be fed with the most liked buttons and their contained features for the machine to learn and generate the best of the bests.

Constructing intelligent systems

AI may employ a design system to build flawless, high-fidelity representations of systems for all data-driven goods, which are loaded with dashboards, tables, and graphs. It can be automated in large part if we give it a set of rules to follow (design system) and wireframes it can understand. Building with a grid system entails arranging prefabricated blocks on a canvas. As tedious and time-consuming as that task is for a human, an AI might complete it in a matter of seconds.

UI created by an AI-powered design system

If the output is a file that can be read by a standard design tool, a human designer can make some creative tweaks, but the bulk of the job will have already been completed. I think it is realistic to assume that people will still be far superior to machines in more imaginative tasks like designing landing pages or developing highly specialized, non-generic software applications.

Creating a working model

Like flowcharts, prototyping has some added complexity. AI will need more time to learn how different acts might have similar effects. The "next" button's behavior might change drastically from one screen to the next. Flows can be too complicated for AI to handle just yet. Hence, it is not feasible to entrust AI with low-level, exact prototyping. There would be too many mistakes to fix, and while it is true that you can train AI through error correction, doing it correctly in a scenario with dozens or even hundreds of interconnected elements would prove challenging. This approach can be successful for a very basic system or app since the obvious flows can be automated at the prototyping phase. Assuming you have not gone password less just yet, AI can duplicate the "forgot my password" process with remarkable accuracy if it is given enough information about how it normally works.

AI-performed micro-interactions

Miniature encounters and transitions animation and interactivity will be one of the last aspects of design to be automated. Twitter pulls to refresh feature is something AI might have developed by accident. The small delight moments that arise through smooth transitions and friendly contacts, on the other hand, have a direct bearing on the material world. With haptic feedback sensations as we activate a choice, or even the length of time it takes to swipe with our finger. Is there a sensation of the window being flung to one side as we swipe it? Many of those wonderful surprises have yet to be found, but I have faith that humans will find them, because even a microsecond's difference in an animation can profoundly affect our emotional response to it.

Automaton artwork: pictorial representation

Artificial intelligence is unlikely to ever be able to replace all the human touches that give a product its unique flair and style. Customers adore these features (especially original artwork) because they help them better remember and relate to the product. Having a distinct visual style for those personal features might help a digital product stand out in a sea of similarly designed offerings. The Internet is rife with popular illustration libraries and premade human body parts. In a matter of seconds, we may have images generated by AI using those libraries. We may instruct it on how to identify feelings based on body orientation. Nevertheless, that's not the same thing. In my opinion, this is not the case with stock illustrations. As they gain notoriety, they more resemble our practice of employing a stock model and telling the world she is a part of our team. In the future, we'll require the services of photographers, models, and artists.

UX in the age of AI

UX designers should instead focus on innovation by letting AI handle the mundane, repetitive tasks. Is this the future? No one can deny the world-altering potential of AI. Even if we do not know for sure if these shifts are for the better, we can at least rest assured that we will spend a lot less time adapting to them. Many of the procedures can be totally or largely taken over by AI soon, and that executing tedious, repetitive jobs should be automated. AI has eliminated the need for us to perform repetitive activities, and we can devote our efforts to further developing the sector through innovation.

6.2   AI Potential in Interaction Design (IxD)

AI and interactive life

Artificial intelligence (AI) and its offshoots, such machine learning and computer vision, are driving change. As a result, it will shape the design landscape and the tools, roles, workflows, and processes of designers. Due to this, we need to stop labeling AI as "artificial" and instead consider it as "augmented intelligence," as well as how we may use it to become our collaborative partner in the creative process. We need to stop thinking of ourselves as humans vs machines and start thinking of ourselves as humans plus machines. Learning about AI will cause us to rethink some of our methods and spark some fresh ideas. Better products, services, and people's quality of life are the goals of AI in design, not user interfaces.

All we hear is that AI is going to take our jobs. Robots are taking over the world! Perhaps, but I wonder if that will occur. What exactly happens to us as designers/ developers? If you type "Is AI going to…" into Google, the autocomplete option will suggest words like "Take over the planet" and "Take our jobs…" Could we be taking artificial intelligence too far? All of these have negative implications if you notice. None of these has any redeeming qualities. Better still, how about "improve our lives" or "rescue the world"! A lot of people are still afraid of AI, and that fear stems mostly from misunderstandings.

Onyx

Computer vision is one of the most promising uses of artificial intelligence. Computer vision is being used in an increasing number of applications, such as enabling the visually impaired to "see" objects, analyzing photos for patterns, and verifying identities through biometric scanning of the face. In fact, AI is now more accurate than humans at categorizing photos, with object detection (or determining an image's subject) having reached 98% accuracy. Many new businesses are developing computer vision-based shopping apps rather than just monitoring infants or driving autonomous vehicles. Envision yourself snapping a photo of a random person on the street whose ensemble strikes your fancy. To those who are brave enough to try it. Odd, perhaps… In a matter of seconds, artificial intelligence will evaluate the picture and present you with a buy button. Currently, users can upload a photo or screenshot to Amazon's app, and the company will conduct a site search for products that are comparable to the ones provided in the user-provided media. Using your phone's camera, this fitness software can track your step count, squat depth, number of push-ups, and more using the power of computer vision and deep learning [4].

Real-time detection

Why don't you say something encouraging and useful? What about a way to make it so that blind individuals can "see" obstacles in their way? They could implant a camera on their breast that uses computer vision to transmit information about their surroundings through sound. There would be distinct sounds for different things, such as cars, bicycles, people, lamp posts, and the location of the pedestrian crossing.

Today's cutting-edge real-time object-detecting system, YOLO, makes this possible. With the help of AI, persons with impairments will have access to previously inaccessible opportunities and experiences. You only look once (YOLO) is a state-of-the-art, real-time object detection system. On a Titan X, it processes images at 40–90 FPS and has a mAP on VOC 2007 of 78.6% and a mAP of 44.0% on COCO test-dev [5].

Personalization

Numerous industries and fields are currently making use of machine learning. For instance, Gmail's spam filters, personalized suggestions (think: any social media feed or Netflix), voice-based interfaces (think: Alexa, Siri, or Google Home), chatbots, computer vision applications like facial recognition, and Google classifying millions of photographs for you to search through. There will be a widespread adoption of AI systems during the next decade, and its impact on society is expected to be far-reaching; some even compare it to the introduction of personal computers.

Pattern recognition and iterative learning are at the heart of each machine learning algorithm. This is already being used in the real world. We get personalization, for instance, when we consider a user's demographic profile, preferences, and activities, such as their location, search terms, and web browsing history (Twitter, Google, Instagram, Netflix). Additionally, we have Amazon advising customers on what to buy, Google Home and Amazon Echo communicating with us, and Waymo traversing the world with autonomous vehicles. Assuming machine learning to be magical is a widespread misconception. Not at all. It is a fantastic piece of equipment [5].

Augmented intelligence

Artificial intelligence (AI) can be thought of as our augmented intelligence, which will increase our efficacy. As cognitive augmentation frees us from tedious, time-consuming duties, we can devote ourselves more fully to innovative product creation. When we finally get this, we can stop worrying and start working with AI because we'll know we can trust it. Artificial intelligence will merely serve as a cognitive enhancer, amplifying and hastening the creative process. *Come on, let's check it out.* Artificial intelligence (AI) can act as a data assistant, helping us collect and analyze information. Making inferences, links, and discoveries from observed data. Sorting through mountains of data collected from user surveys, interviews, observations, audio recordings, and video recordings. Artificial intelligence (AI) may learn from user feedback to determine the optimal design pattern.

AB testing

With the help of AI, designers and developers will be able to rapidly test new designs. Consider instantaneous eye-tracking heat-map generation for thousands of users, the results of multivariate and A/B tests, and the testing of emotional responses using facial recognition. With this streamlined procedure, we can gain useful insights much more quickly [6].

1. **Incorporate variations**

You plug in your link and play around with various iterations of your headline, body copy, and CTA. The number of permutations resulting from these changes can be enormous. Each of the headings, body text, and call-to-action will be recognized by our AI.

- It will provide you with alternate suggestions mechanically.
- The AI will detect the moment of your conversion without any input from you.

2. **Combine**

Incorporate our JavaScript code snippet into your site or use our plug-in for WordPress/Wix.

- Smaller than 1 KB in size, our JavaScript code sample is extremely efficient.
- There will be no noticeable slowdown in performance while the tests are being run.

3. **Proceed with the first experiment**

Then, using a multiarmed bandit strategy, we test out a sample of the possible permutations.

- The multiarmed bandit algorithm makes it highly improbable that your overall conversion rate will decrease as you do the trial.
- By repeatedly displaying the most promising variant, the multiarmed bandit algorithm ensures that no conversions are lost during testing.

4. **Use AI to find the best combination**

Applying artificial intelligence to the task of finding the optimal combination.

- Once statistical significance is reached in an experiment, the optimal variations are extracted and recombined with a new set of parameters using an evolutionary algorithm.
- Instead of trying every potential combination, we may quickly determine the optimum variant by utilizing an evolutionary algorithm.

You can start seeing the results of A/B testing sooner and without needing a large volume of traffic thanks to the fact that we execute our studies in batches. To determine the optimal conversion-boosting combo, we repeatedly execute experiments.

Choosing best UI

The use of artificial intelligence (AI) to find the best user interface (UI) design would be greatly facilitated if, early in the product design process, we could swiftly generate a wide variety of UIs for testing. In a fraction of the time, it used to take, we could now check designs against established norms for usability.

Let us pretend we are developing a business-to-business dashboard. We then feed this design into an AI-powered UI analysis tool. We perform a battery of tests based on established criteria for usability, accessibility, and interaction design. One of the test results displayed is seen here. This is state-of-the-art AI and computer vision technology. It involves spotting trends, conducting analyses, and highlighting problems that need fixing.

It appears that there are problems with usability and accessibility in this design. We can make changes and rerun the tests until we have an optimal design. Artificial intelligence allows us to complete this task in a fraction of the time it once required. Also, in a matter of minutes, we can produce complete design systems, component libraries, and even complete style guides. To ensure that all members of the team always have access to the most recent and most stable versions of the system's components, our AI tools may build the design system, replete with code that's ready to be used. (I'd be willing to bet that a few different businesses are simultaneously developing this very tool). Using computer vision, a branch of AI that creates complete user interfaces and code from sketches, Airbnb now has an experimental system in place. It's software that can take a library of premade UI elements and use them to produce a working prototype. Everything is thrown together hastily and set to go. OK, here it is.

Sketch2Code

Microsoft's computer vision system can help product teams iterate designs more quickly. By leveraging AI and computer vision, Sketch2Code can take a user's sketch and turn it into a fully functional HTML prototype. Whiteboards provide for easy collaboration among the product team. Simply snap a photo and run it through an AI-powered image recognition system to get the corresponding code. Sketch code is a web-based application which converts hand-drawn diagrams into HTML code base. Once you have drawn wireframes on a whiteboard or a sheet of paper, take a picture using the web app. The web app sends the information to the AI service. Then the AI service runs those images against a prebuilt AI model and then creates a HTML code base followed by a resulting app.

At the core of the system is a custom vision of the prediction model which is trained with a set of hand-reading images from several different people and the model is taught to identify the basic HTML elements like buttons, labels, and text boxes. Once the model is ready, it can predict one of those trained elements present on a given image. On the other hand, handwritten recognition service guesses what's reading inside these elements. With these two pieces of information and geometry, the HTML code is built. There Is a Requirement for a User Interface An important part of the design process is the exchange of ideas, which typically begins at a whiteboard. Once a concept has been sketched, it is typically photographed and then manually converted into an HTML wireframe. This is a time-consuming task that slows down the design procedure. Microsoft's Cognitive Services is home to the company's own Computer Vision Service. This service's model, which has been trained using millions of photos, can detect a wide variety of items. Specifically, we need to develop a unique model and educate it using pictures of hand-drawn UI components like text boxes, buttons, and combo boxes. We can now train our own models and have them do object identification thanks to the Custom Vision

Service. Once HTML items have been identified, we employ the Computer Vision Service's text recognition features to pull any handwritten text from the design. Using these two pieces of data in tandem, we can build the HTML code for the design's many components. By looking at where these pieces are located, we can extrapolate the design's layout and write the HTML code. Machine learning automates routine tasks, freeing up human engineers and designers to focus on conceptualizing and implementing new features. Artificial intelligence (AI) is complementary to human creativity rather than a threat to it. We will be shaping both the intake and the product, after all.

Automated UI mockups

Artificial intelligence systems can rapidly assemble user interface designs, which will allow us to conduct simulations on high-fidelity mockups. The AI will allow us to get things into people's hands rapidly to test and learn, even when they are not yet complete. The next step is fast iteration, during which design optimization is performed in real time. Because of this, we would be able to fully explore our creative potential.

By incorporating ML into our products, we can improve UX by using AI-driven recommendations for extreme customization. This form of anticipatory design allows us to foresee a user's likely tastes in media and products. We're getting close to being able to do so, which will allow us to provide personalized digital experiences to everyone. Media, travel, news, social media, e-commerce, and more will all benefit from the improved user experience brought about by hyper-personalization powered by AI. There is a growing availability of resources that facilitate the use of AI in product development.

6.3   UX Writing (UXw) for AI

The significance of UX writing in the context of machine learning. How does it aid users in constructing accurate cognitive frameworks?

The primary objective of UX writing is to elucidate the functioning of machine learning to individuals who incorporate it into their everyday activities. The operational mechanisms of machine learning (ML) within the applications and tools we employ, as well as the user experience (UX) writing, may often go unnoticed. UX writing explicitly informs users about the expected outcomes of their actions, delineating the specific functionalities of the underlying algorithm, such as performing x, y, and z operations, and ultimately yielding a predetermined output. This statement elucidates the importance of clearly communicating the rationale behind decision-making processes, including the information considered and the signals conveyed. Furthermore, it emphasizes the significance of managing expectations regarding the outcomes, and highlights the potential applicability of these results to individuals. Additionally, it raises the question of how individuals can effect change in future decision-making endeavors, provided that the process is executed correctly. In essence, the purpose is to establish clear parameters regarding the operational functionality of this apparatus, delineate the anticipated outcomes for users, and elucidate the means by which they can modify said outcomes. Machine learning involves the process of generating predictions, which are subsequently evaluated by users to determine their accuracy. This feedback loop facilitates the machine's learning and enables it to acquire additional knowledge over time. For instance, in the event that an algorithm proffers a movie recommendation and the user declines by stating "no thank you," it can be inferred that the prediction was inaccurate. There exists an additional aspect wherein an application proposes to Travis the option of dining out on Thursday at a sushi establishment that offers al fresco seating and discounted prices during specific hours. However, Travis declined the suggestion, expressing a lack of interest in pursuing such an activity. Subsequently, the application presents Travis with a selection of options to indicate the reasons why the prediction did not align with his preferences. The manner in which these options are formulated in a detailed manner will provide Travis with insights into the underlying workings of the algorithm. In this particular instance, the factors contributing to the situation may include the time of day, the evening, the meal being dinner, the specific cuisine being sushi, the day of the week being Thursday, the presence of happy hour promotions, and potentially inclement weather conditions such as rain. Consequently, the availability of outdoor seating may not be ideal for your preferences. The provision of feedback and the organization of writing aids Travis in comprehending and constructing a cognitive framework for understanding the functioning of these predictions, as well as Travis' responsibility to enhance them over time.

The concept of mental models within the field of machine learning

A mental model refers to the cognitive representation of the underlying mechanisms and processes that govern the functioning of an object or a phenomenon. A conceptual framework for a bicycle entails the presence of two pedals and the act of steering the handle, which enables the bicycle to move forward on the ground. It is important to note that the speed at which one pedals directly influences the rate of acceleration. An illustrative mental framework for a video recommendation application entails the process of viewing videos, subsequently prompting the application to suggest additional videos that bear resemblance to those previously viewed by the user. There is a possibility of carryover if one of the sequences deviates, resulting in the accumulation of watched videos in a stack commonly referred to as a "start pile" in colloquial terms. The metal model serves as a means to illustrate the impact of the starting pile on all other elements. If the initial stack of items is undisclosed or remains pertinent to your subsequent suggestions, may you please reset your initial stack, or alternatively, reset all elements involved? The inquiry pertains to understanding the methodology employed in order to achieve desired outcomes, which constitutes the underlying mental framework.

In order to elucidate the concept of mental models in the context of Google Flights, the individual in question highlights the feedback received by both herself and the Google Flights team from users. It has been observed that users tend to associate the increase in prices displayed by Google Flights with their frequency of interactions with the platform's interface. This correlation forms the basis of the users' mental model. The user's mental model regarding Google Flight prices is typically as follows. According to Pinto, the current conceptualization of Google Flights is flawed. She discusses the approach taken by Google Flights to address this issue, aiming to enhance its clarity by adopting a format similar to that of a weather update. The individual in question refers to it as a hyper-personalized cognitive framework, which they endeavored to transform into a weather update by providing suggestions such as, "Historically, flight prices for your specific booking scheme tend to be elevated during this time period, and they are likely to increase further as the dates approach." This mental model is significantly more comprehensible and feasible, making it easier for individuals to evaluate it objectively rather than perceiving the predictions as solely applicable to themselves. Depersonalization refers to the process of making predictions more predictable, thereby enhancing user trust in the reliability of predictions made by Google Flights. This is where the field of UX writing assumes a crucial role.

These system workflows and user interactions possess an additional layer. There exists a delicate equilibrium between excessively elaborating on concepts and excessively hedging the users' forecasts with an excessive number of confidence levels. As an illustration, there exists a 90% level of confidence that the price will experience an increase of 3 USD within the upcoming two-day period. The assistance provided to the user is not deemed beneficial. The user inquiries about the remaining duration until the price increase occurs. In summary, UX writers serve as intermediaries by facilitating the translation of machine-generated recommendations to users, while also ensuring that users comprehend these predictions at a more abstract level in order to inform their decision-making processes.

Finding the optimal equilibrium between simplicity, accuracy, and complexity in comprehending Google flight recommendations or similar platforms is a crucial consideration

In instances where users possess a limited understanding of recommendations, primarily stemming from an excessive number of explicit confidence levels, there exists a potential for overreliance and subsequent negative consequences. Flight tickets are of relatively low concern to customers, as they possess the ability to cancel their booking if it fails to meet their requirements. However, it encompasses user preferences, aversions, trust, and other dimensions of customer behavior, which hold significant importance for companies such as Google Flights. The objective is to provide users with clear and comprehensive explanations, enabling them to delve further into the intricacies of the data. This includes ensuring that the data is representative, segmented, cleansed, and analyzed according to their specific requirements, thereby facilitating effective communication. Another objective is to effectively convey this information using graphs, illustrations, visualizations, interactive methods, and written explanations, which is the responsibility of a UX writer.

What level of difficulty is involved in translating machine language to human language?

In contemporary times, individuals have developed a greater familiarity with probabilistic applications, such as recommendation systems, within the context of the digital realm. Consequently, this enhanced familiarity has contributed to a perceived ease in engaging with such systems. The act of providing recommendations, whether accurate or erroneous, carries implications for the establishment and maintenance of trust. Empathy is a crucial factor in obtaining a deeper understanding of users' decision-making processes. However, Roxxane presents a contrasting perspective regarding the concept of empathy, specifically in relation to rectifying mistakes. The individual endeavors to contemplate the malfunctioning of a product or feature, along with devising solutions for each encountered "error," rather than expressing contentment regarding the potential positive outcomes in a prognostic manner. Consider a scenario where a machine learning assistant within a fitness application engages with the user by stating, "Greetings, I am your designated fitness instructor." In what manner may I be of assistance to you? In response, a cordial greeting is extended, expressing enthusiasm towards the prospect of collaborating. The virtual assistant displays an "error" message, indicating its inability to comprehend the input provided. The reason for this is the necessity for the tone of the voice to align with that of the trained assistant, as it may be unfamiliar to the user. Hence, error messages serve as an effective means of fostering trust in such circumstances. The error message should inform the user that it will provide assistance during challenging situations and offer suggestions for troubleshooting, rather than relying on unhelpful error messages.

Each product or application possesses a distinct tone or voice. A UX writer typically employs a tone map as a means of providing guidance to users. A tone map typically functions as a collection of various vocal tones, encompassing emotions such as happiness, enthusiasm, neutrality, concern, and comfort, among others. The objective of this study is to analyze the correlation between user situations and the tone of the assistant.

The user experience (UX) of artificial intelligence (AI)

The primary concern revolves around ensuring that designers possess a comprehensive understanding of the appropriate contextual utilization of machine learning. Due to the considerable costs associated with machine learning (ML), the substantial volume of data required is remarkably high, while the level of expertise demanded from engineers is exceedingly difficult to fathom. It is imperative to cultivate a heightened sense of awareness among individuals, encouraging them to exercise caution and engage in critical thinking from multiple perspectives prior to asserting their proficiency in machine learning and its application in any given scenario. How can the utilization of machine learning technology be directed towards appropriate objectives? Assisting individuals, ensuring user satisfaction, and streamlining processes relevant to human interests.

Final words

We as interaction designers need to use the ever-expanding and never-ending capacity of machine learning to create amazing user-interactive products. We must also use this technology to design for all. A call for inclusive design is primary with the interaction design field.

References

1. Malewicz, M. 2020. *Will AI take over UX?* https://uxdesign.cc/will-ai-take-over-ux-ea164a 2ed39f.
2. *Machine learning and its different types of algorithms.* https://dribbble.com/shots/17169353-Machine-Learning-and-its-different-types-of-Algorithms.
3. Sharma, K. 2019. *Neumorphism (Soft UI) in User interface design.* https://uxplanet.org/neumor phism-in-user-interface-tutorial-c353698ac5c0.
4. Onyx. 2020. *Onyx—The world's smartest digital trainer.* https://www.youtube.com/watch?v= DN7lRduwyG8.
5. Philips, M. 2020. *AI and design: Why AI is your creative partner.* https://uxdesign.cc/ai-and-des ign-ai-is-your-creative-partner-cb035b8ef107.
6. ABTesting. *Optimize your landing page for conversions using our AI A/B testing software.* https://abtesting.ai/.

Conclusion

In conclusion, AI is rapidly changing the landscape of design, offering designers new and innovative tools and techniques to create more engaging and impactful experiences for users. As the technology continues to evolve, designers who are able to harness the power of AI will have a distinct advantage in the marketplace. This book on AI for Designers by Md Haseen Akhtar and Janakarajan Ramkumar is a valuable resource for designers who are looking to expand their knowledge and skills in this exciting field. The book covers the fundamentals of AI, explains its applications in design, and provides practical guidance on how to work effectively with AI tools and technologies. Throughout the book, we the authors emphasized the importance of embracing AI as a tool for generating new and innovative ideas, rather than simply automating repetitive tasks. He shows how designers can use AI to create generative art, unique interfaces, and cocreate with users. As we look to the future of design, it is clear that AI will play an increasingly important role. By staying up-to-date with the latest developments and techniques, designers can leverage this powerful technology to create more impactful and meaningful experiences for users. We highly recommend this book to anyone who is interested in learning about the creative potential of AI in design. With its clear and concise explanations, practical examples, and emphasis on creativity, this book is a must-read for designers who are looking to stay ahead of the curve in the rapidly evolving world of AI.

The competition between Bing, Bard, and ChatGPT is helping to drive innovation and improve the quality of human-centered conversational AI in several ways. As these companies compete to create the most advanced AI models, they are pushing the boundaries of what is currently possible in natural language processing and conversational AI. This is leading to the development of new techniques and approaches that can be applied more broadly to improve the quality of AI-driven conversations. While these companies are competitors, they also have an incentive to work together to advance the field of conversational AI. This collaboration can take many forms, such as shared research and development projects or the creation of open-source tools and resources. Ultimately, the goal of these companies is to create conversational AI

© The Editor(s) (if applicable) and The Author(s), under exclusive license
to Springer Nature Singapore Pte Ltd. 2024
M. H. Akhtar and J. Ramkumar, *AI for Designers*,
https://doi.org/10.1007/978-981-99-6897-8

systems that are useful and engaging for users. By competing with one another, they are constantly working to improve the user experience and make these systems more intuitive, responsive, and effective. As competition increases, companies are under pressure to create more efficient and cost-effective AI models. This can lead to the development of new techniques that improve the speed and accuracy of AI-driven conversations, while also reducing the resources required to run these systems. Overall, the competition between Bing, Bard, and ChatGPT is helping to drive innovation and improve the quality of human-centered conversational AI in a variety of ways. As these companies continue to push the boundaries of what is possible, we can expect to see continued improvements in the quality and effectiveness of conversational AI systems.

Human-centered artificial intelligence (AI) is likely to be changed in the future by a number of things, such as technological progress, ethical concerns, and changes in society. As AI becomes more common in our daily lives, we will pay more attention to ethical issues like fairness, transparency, and accountability. To do this, AI systems will need to be carefully designed and built so that they are in line with human values and goals. Natural language processing is one area where AI is likely to make big strides. This will make it easier for people and machines to work together in a way that feels more natural and intuitive. This will be especially important in areas like customer service and health care. As AI gets better at figuring out how people act and what they like, we can expect to see more personalized experiences in a lot of different areas, for example, personalized recommendations for products and services, customized healthcare treatments, and customized education plans. New technologies like the Internet of things (IoT) and blockchain are likely to be combined with AI. This will make it easier for different systems and devices to work together and open up new ways for innovation and growth. Overall, the future of human-centered AI is likely to be marked by a focus on ethics, more personalization and natural language processing, and integration with other new technologies. But it's important to keep in mind that AI development is a complicated and quickly changing field, so it's hard to know how things will go in the coming years. Talk about Bing, Bard, and ChatGPT.

"It is not the end of this discussion, it is just a starter to the new wave of AI and its implications on human race"

–Authors

Printed in the United States
by Baker & Taylor Publisher Services